CAMBRIDGE MONOGRAPHS IN
EXPERIMENTAL BIOLOGY
No. 17

AN ECOLOGICAL APPROACH TO ACANTHOCEPHALAN PHYSIOLOGY

THE SERIES

1 V. B. WIGGLESWORTH. The Physiology of Insect Metamorphosis

2 G. H. BEALE. The Genetics of *Paramecium aurelia*

3 G. V. T. MATTHEWS. Bird Navigation. 2nd edition

4 A. D. LEES. The Physiology of Diapause in Arthropods

5 E. B. EDNEY. The Water Relations of Terrestrial Arthropods

6 LILIAN E. HAWKER. The Physiology of Reproduction in Fungi

7 R. A. BEATTY. Parthogenesis and Polyploidy in Mammalian Development

8 G. HOYLE. Comparative Physiology of the Nervous Control of Muscular Contraction

9 J. W. S. PRINGLE. Insect Flight

10 D. B. CARLISLE and SIR FRANCIS KNOWLES. Endocrine Control in Crustaceans

11 DAVID D. DAVIES. Intermediary Metabolism in Plants

12 W. H. THORPE. Bird-song

13 JANET E. HARKER. The Physiology of Diurnal Rhythms

14 J. E. TREHERNE. The Neurochemistry of Arthropods

15 R. N. ROBERTSON. Protons, Electrons, Phosphorylation and Active Transport

16 GEORGE SALT. The Cellular Defence Reactions of Insects

17 D. W. T. CROMPTON. An Ecological Approach to Acanthocephalan Physiology

AN ECOLOGICAL APPROACH TO ACANTHOCEPHALAN PHYSIOLOGY

BY

D. W. T. CROMPTON

Lecturer in Parasitology,
The Molteno Institute,
University of Cambridge
(Fellow of Sidney Sussex College)

CAMBRIDGE
AT THE UNIVERSITY PRESS
1970

CAMBRIDGE UNIVERSITY PRESS
Cambridge, New York, Melbourne, Madrid, Cape Town, Singapore, São Paulo, Delhi

Cambridge University Press
The Edinburgh Building, Cambridge CB2 8RU, UK

Published in the United States of America by Cambridge University Press, New York

www.cambridge.org
Information on this title: www.cambridge.org/9780521104708

First published 1970
This digitally printed version 2009

A catalogue record for this publication is available from the British Library

Library of Congress Catalogue Card Number: 73–118062

ISBN 978-0-521-07939-6 hardback
ISBN 978-0-521-10470-8 paperback

7

CONTENTS

ACKNOWLEDGEMENTS

IN writing this monograph, I have relied on many people, especially Dr George Salt, for guidance and assistance. Drs Penelope Butterworth and P. J. Whitfield have generously allowed me to refer to their unpublished work. Miss Susan Arnold, Mr D. Barnard and Drs A. P. M. Lockwood, B. A. Newton, Susan Rotheram, P. Tate and P. F. V. Ward have helped me in many ways. Mr Paul Knights of Cambridge Scientific Instruments Ltd has provided the electron micrograph for plate 2B. Miss Stephanie Hamer has skilfully prepared the typescript. To all these helpers, I wish to express my gratitude.

I also thank the following publishers and copyright owners: Academic Press for permission to use plate 2A and fig. 17 from *Experimental Parasitology*, and fig. 14*b* from *Advances in Parasitology*; Akademische Verlagsgesellschaft Geest and Portig, K-G, Leipzig, for fig. 14*a* from *Bronn's Klassen und Ordnungen des Tierreichs*, 4, *Acanthocephala* by A. Meyer; the American Institute of Nutrition for figs. 5*c* and *d* from the *Journal of Nutrition*; the American Society of Parasitologists for figs. 4*d* and *e*, 9, 10, 14*c*, *d* and *f*, 22*a*, *c*, *d*, *e*, *h* and 24*a* from the *Journal of Parasitology*; Cambridge University Press for figs. 3*b*, 6, 11, 12, 13, 18*e* and 21 from *Parasitology*; the Company of Biologists Ltd for figs. 5*a*, *b*, *e*, *f*, *g*, *h* and 8 from the *Journal of Experimental Biology*; John Wiley and Sons Inc. for fig. 1 from *Systema Helminthum. V. Acanthocephala* by S. Yamaguti; Gustav Fischer Verlag for fig. 14*e* from 'Urhautzell, Hautbahn und plasmodiale Entwicklung der Larve von *Neoechinorhynchus rutili* (Acanthocephala)' by A. Meyer in *Zoologische Jahrbücher, Anatomie und Ontogenie der Tiere*, **53**, 103–26; Springer Verlag for fig. 24*b* from 'The occurrence of *Pallisentis basiri* Farooqi (Acanthocephala) in the liver of *Trichogaster chuna*' by R. Hasan and S. Z. Qasim in *Zeitschrift für Parasitenkunde*, **20**, 152–56. My acknowledgement to the authors concerned has been made in the text.

Cambridge,
December, 1969

D.W.T.C.

CHAPTER I

General Introduction

ALTHOUGH most groups of helminths were known to the ancients, the Acanthocephala remained unrecognized until the late 17th century when individuals parasitic in eels were described independently by Redi (1684) and Leeuwenhoek (1692). Redi stated that in all types of eels he often found tiny white worms which were tightly fixed by their extremities to the intestinal wall. He also reported, 'the worms possess a proboscis which is equipped with hooks and which the worms are accustomed to send out and draw in'. Leeuwenhoek wrote in a letter on reproduction in eels, sent to The Royal Society on 16 September 1692, that he had noticed two types of Acanthocephala in the eel's intestine. His first impression was that the worms were young eels of the next generation, but after dissecting one he changed his mind. One type of worm was reported to be small and red and to contain a great number of little worms. He was probably describing mature female acanthocephalan worms, which may have been specimens of *Pomphorhynchus laevis*. The other type was described as white with tiny joints, which is at first suggestive of a cestode; but many acanthocephalans will contract in an irregular manner on interference and Leeuwenhoek not only described but also figured an unmistakable acanthocephalan proboscis belonging to this parasite. He wrote that the heads of these worms were sticking to the intestine and, when he examined them under the microscope, it was with amazement that he saw numerous 'hooky' parts. He also commented that he saw, when he could manage to wrench them away unharmed from the intestines, that they retracted this 'hooky' part inside their bodies. It is not unlikely that these worms, and those found by Redi, were specimens of *Acanthocephalus anguillae*.

Since these early descriptions, approximately 650 species of acanthocephalans have been identified from all over the world, and, as Redi and Leeuwenhoek observed, the retractile proboscis is the most obvious and characteristic feature of adult worms (fig. 1). In addition, acanthocephalans are diœcious, pseudocoelomate worms without an alimentary tract at any stage of their development. They are endoparasitic throughout their indirect life cycles and the structure of the body wall and

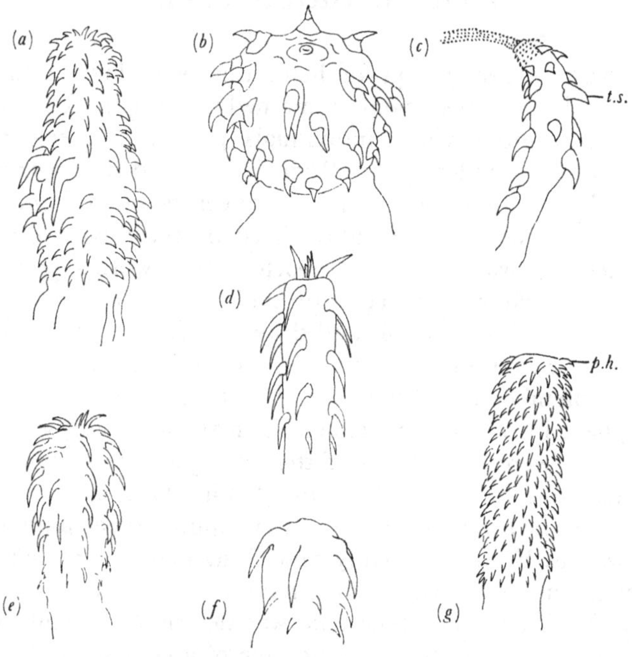

Fig. 1. The proboscides of seven species of Acanthocephala. (*a*) *Corynosoma turbidum*; (*b*) *Macracanthorhynchus hirudinaceus*; (*c*) *Rhadinorhynchus horridum*; (*d*) *Acanthocephalus tenuirostris*; (*e*) *Acanthocephalus anguillae*; (*f*) *Neoechinorhynchus rutili*; (*g*) *Echinorhynchus gadi*. (After Yamaguti, 1963; figs. 6, 134, 150, 194, 328, 416 and 630) *p.h.*, proboscis hook; *t.s.*, trunk spine.

the anatomy of the reproductive systems are peculiar to the group. Acanthocephalans appear to be related to the aschelminths, but it is probably most satisfactory to consider them as a separate phylum. In this book, the classifications of Hyman (1951) and Rothschild (1961) have been adopted for the acanthocephalans and their hosts; the letter A denotes order Archiacanthocephala, P denotes Palaeacanthocephala and E denotes Eocanthocephala on the figures and tables.

Mature male acanthocephalans are usually smaller than mature females of the same species. The majority of species are about 1 to 2 cm long, but some are much longer, notably female *Macracanthorhynchus hirudinaceus* from pigs (Kates, 1944) and female *Nephridiacanthus longissimus* from aardvarks (Golvan, 1962) measuring 45 cm and 93 cm respectively. The worms are unsegmented, but superficial annulations are often present and may give the impression of segmentation.

Various aspects of the anatomy of adult acanthocephalans are shown diagrammatically in figs. 2 and 3. The term praesoma was coined by Rauther (1930) for the proboscis sheath, the lemnisci and all the structures involved in the function of the proboscis as an organ of attachment. The remaining structures form the metasoma, which includes all the body wall situated posterior to the partition. This partition ensures the flow of fluid between the lemnisci and the wall of the neck (figs. 2*a*; 3*a*) and prevents the movement of fluid between the metasomal body wall and the lemnisci. The division of the acanthocephalan body into the praesoma and metasoma is a matter of descriptive convenience only; the regions are interdependent and integrated and cannot function alone.

The proboscis is equipped with a variety of hard, sharp hooks and the anterior portion of the body wall often bears trunk spines which also assist in the attachment of the parasite to its host's intestinal wall. The proboscis is withdrawn into its sheath by contraction of the proboscis retractor muscles and is everted within a few seconds by a hydrostatic system (Hammond, 1966*a*, *b*). After withdrawal of the proboscis, the neck may also be withdrawn when the neck retractor muscles contract. The compact folding of the worm, resulting from contractions of this type, may be observed *in vitro*. It is unlikely to occur in the intestine where the worms must maintain contact with their hosts by means of either their proboscides or trunk spines to avoid being expelled by peristalsis. Contraction of the retractor muscles by an attached worm will pull the metasoma or trunk against the intestinal wall and away from abrasive or solid material in the lumen. One layer of circular muscles and one of longitudinal muscles are found beneath the body wall. All

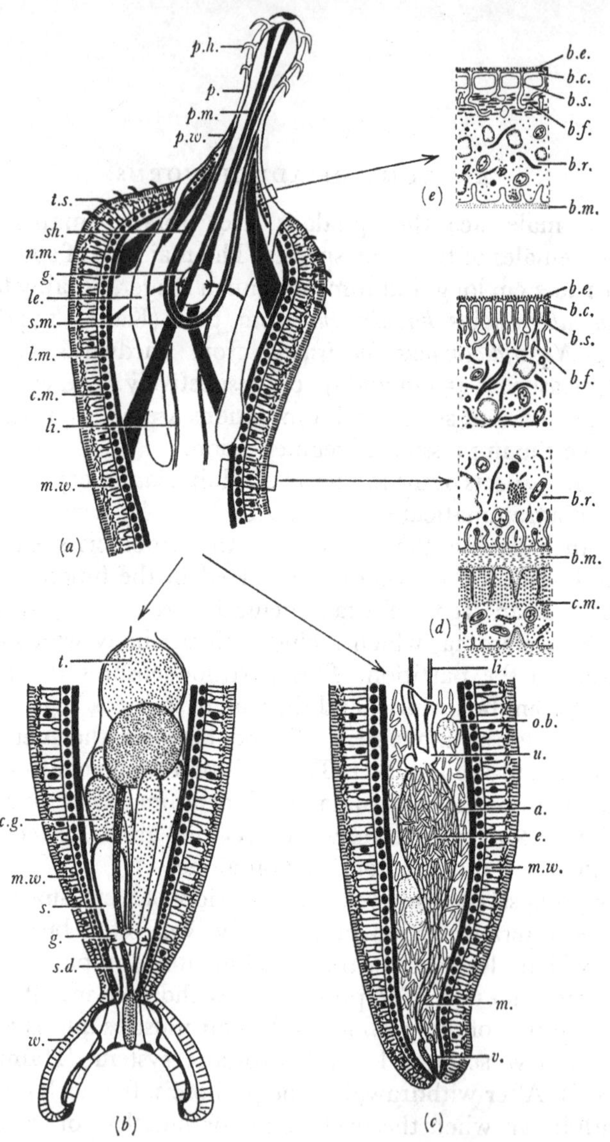

Fig. 2. Diagrammatic representations of the anatomy of an adult acanthocephalan worm. (*a*) Structures common to both sexes; (*b*) male reproductive organs; (*c*) female reproductive organs; (*d*) metasomal body wall; (*e*) praesomal body wall. *a.*, anterior uterus; *b.c.*, cuticle; *b.e.*, epicuticle; *b.f.*, felt layer; *b.m.*, basement membrane; *b.r.*, radial layer; *b.s.*, striped layer; *c.g.*, cement gland; *c.m.*, circular muscle; *e.*, eggs; *g.*, ganglion; *le.*, lemniscus; *li.*, ligament; *l.m.*, longitudinal muscle; *m.*, muscular lower uterus; *m.w.*, metasomal wall; *n.m.*, neck retractor muscle; *o.b.*, ovarian ball; *p.*, proboscis; *p.h.*, proboscis hook; *p.m.*, proboscis retractor muscle; *p.w.*, praesomal wall; *s.*, Saefftigen's pouch; *sh.*, proboscis sheath; *s.d.*, sperm duct; *s.m.*, proboscis sheath retractor muscle; *t.*, testis; *t.s.*, trunk spine; *u.*, uterine bell; *v.*, vagina; *w.*, bursa.

acanthocephalan muscle so far examined in detail has been found to consist of a contractile and a non-contractile portion (fig. 2*d*).

Simplified representations of the reproductive systems of male and female worms are shown in fig. 2*b* and *c*. The male system, which is completely enclosed in the ligament, consists of a copulatory bursa, several cement glands, a pair of testes and a sperm duct. The female system, with which the ligament is only partly associated, is composed of the uterine bell where mature and immature eggs are sorted, and the uterus from which mature eggs are released. The eggs develop in the body cavity of the female and are nourished from the pseudocoelomic fluid, which surrounds the reproductive organs of both male and female worms. The male system includes a complex of ganglia connected to sense organs located on the copulatory bursa (fig. 2*b*). The greater part of the nervous system in both male and female worms is the ganglion situated in the proboscis sheath through which the lateral nerves emerge (fig. 2*a*).

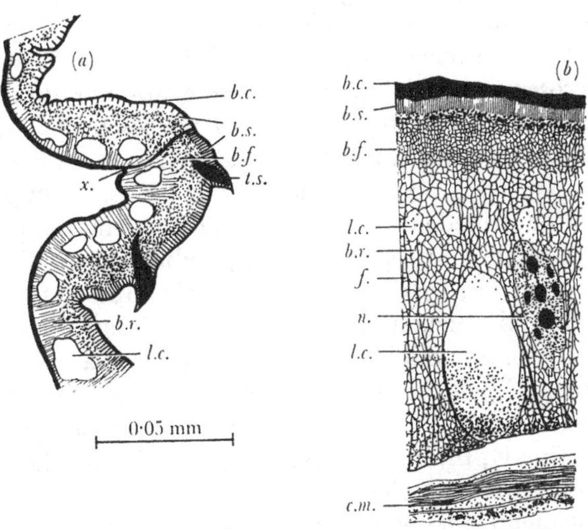

Fig. 3. (*a*) The junction of the praesomal and metasomal body wall of *Polymorphus minutus*; (*b*) the metasomal body wall of *P. minutus* (after Crompton, 1963; fig. 12) *b.c.*, cuticle; *b.f.*, felt layer; *b.r.*, radial layer; *b.s.*, striped layer; *c.m.*, circular muscle; *f.*, fibre; *l.c.*, lacunar channel; *n.*, nucleus; *t.s.*, trunk spine; *x.*, partition between praesoma and metasoma.

Close attention must be paid to the morphology of the adult

body wall; it has become specialized for feeding, metabolism excretion and protection. Furthermore, the body wall is a component of the hydrostatic skeleton and the tissue in which the hooks and spines develop. In fact, a study of the physiology of the acanthocephalan body wall would involve most aspects of acanthocephalan physiology.

The body wall consists of at least five layers. The outermost layer is a thin epicuticle which appears to contain mucopolysaccharide (Crompton, 1963; Wright and Lumsden, 1968). The next layer is the tough cuticle which is penetrated by many pores. Their existence was postulated from studies with the light microscope, but they were not demonstrated until the body wall was examined with the electron microscope. The pores lead into the canals and ducts of the striped layer which blends into the underlying fibrous felt layer. The fibres, the cuticle, the matrix of the striped layer and all skeletal elements are probably composed of protein, or lipoprotein, stabilized by disulphide linkages (Mueller, 1929; Monné, 1959; Crompton, 1963). The innermost and thickest layer is the radial layer containing nuclei, ribosomes, mitochondria, folded plasma membranes, glycogen and lipids, but showing no evidence of cell walls. This layer is clearly the location of much synthetic and metabolic activity. Finally, the radial layer is separated from the circular and longitudinal muscles of the worm by a relatively thick basement membrane of connective tissue. These aspects of the body wall are illustrated and supplemented by fig. 3*b* and plates 1 and 2.

So far, the ultrastructure of the body wall has been examined in detail for only three palaeacanthocephalans, *Polymorphus minutus* (Crompton and Lee, 1965), *Pomphorhynchus laevis* (Stranack, Woodhouse and Griffin, 1966) and *Acanthocephalus ranae* (Hammond, 1967*a*), and one archiacanthocephalan, *Moniliformis dubius* (Nicholas and Mercer, 1965). In spite of the differences in nomenclature and interpretation of electron micrographs discussed by these authors, it emerges that the adult body wall is a uniform, syncytial tissue. More comprehensive accounts of acanthocephalan morphology are to be found in the works of Meyer (1933), Hyman (1951) and Petroschenko (1956; 1958).

All the known life cycles of acanthocephalans involve an arthropod as an intermediate host in which development occurs, and a vertebrate as a final host in which sexual reproduction takes place (table 1). Mature eggs are released from the body cavities of female worms into the intestine of the final host, from which they are discharged with faeces on to soil or into water. It may be inferred from table 1 that the eggs of many archiacanthocephalans will be deposited on soil and eaten by terrestrial insects, while those of many palae- and eoacanthocephalans will fall into water and be eaten by crustaceans. After ingestion by the correct intermediate host, the egg hatches to liberate the acanthor larva, which moves out of the intestinal lumen into the host's haemocoele and there develops into the acanthella. In some parasites the acanthor is retarded in the intestinal tissues during its migration into the haemocoele and, consequently, the change to the acanthella is delayed. When development of the acanthella is finished, it encysts and becomes a resting stage called the cystacanth which remains dormant until its host is eaten by the correct final host or by a transport host. When a suitable final host ingests a cystacanth, the parasite is activated and an immature worm becomes established in the appropriate place in the intestine. Alternatively, if a suitable transport host swallows a cystacanth, the parasite everts its proboscis, migrates through the intestinal wall and becomes encapsulated in the abdominal tissues where it remains until the transport host is eaten by the final host. The literature contains many references to transport hosts, but these have been omitted from table 1 because direct evidence of the necessity for these hosts in most acanthocephalan life cycles is not available. One exception is provided by the gasteropod, *Campeloma rufum*, which could have been cited as a transport host for the eoacanthocephalan, *Neoechinorhynchus emydis*, but now appears to be as essential a host as the ostracod, *Cypria maculata* (Hopp, 1954). The evidence indicates that the parasite actually grows in *C. rufum*, which is better defined, therefore, as a second intermediate host. Transport hosts are usually assumed to harbour and support an acanthocephalan without any growth occurring. Of course, some final hosts, such as seals, are unlikely to eat

TABLE 1. *Some Acanthocephala of which the life cycles have been determined*

Species	Intermediate hosts	Final hosts	References
Archiacanthocephala			
Macracanthorhynchus hirudinaceus	*Cotinus nitida* *Diloboderus abderus* *Melolontha vulgaris* *Phyllophaga rugosa* (coleopteran larvae)	Wild and domestic pigs	Wolffhügel (1908) Kates (1944)
M. ingens	*Phyllophaga crinita* *P. hirtiventris* *Ligyrus* spp. (coleopteran larvae)	*Procyon lotor* (raccoons)	Moore (1946*b*)
Mediorhynchus grandis	*Arphia luetola* *Chortophaga viridifasciatus australior* *Orphuella pelidna* *Schistocerca americana* (adult orthopteroids)	*Turdus migratorius* *Quiscalus quiscala* and other birds	Moore (1962)
Moniliformis clarki	*Ceuthophilus utahenis* (camel cricket)	*Peromyscus maniculatus sonoriensis* (deer mouse)	Crook and Grundmann (1964)
M. dubius	*Periplaneta americana* (cockroach)	Wild and domestic rats	Moore (1946*a*)
Prosthenorchis elegans	*Blatella germanica* (cockroach) *Lasioderma serricorne* *Stegobium paniceum* (beetles)	*Saimiri sciurea* (squirrel monkey) and other primates	Stunkard (1965)
P. spirula	*Blatella germanica*	*Saimiri sciurea*	Yamaguti (1963)

Palaeacanthocephala			
Acanthocephalus nanus	*Asellus aquaticus* (fresh-water isopod)	*Rana rugosa* *Diemyctylus pyrrhogaster* (amphibians)	Yamaguti (1935)
Echinorhynchus truttae	*Gammarus pulex* (fresh-water amphipod)	*Salmo trutta* (trout)	Awachie (1966)
Filicollis anatis	*Asellus aquaticus*	Anatidae (ducks, geese and swans)	Styczyńska (1958)
Leptorhynchoides thecatus	*Hyalella azteca* (fresh-water amphipod)	*Ambloplites rupestris* *Huro salmoides* (fresh-water fish)	DeGiusti (1949*a*)
Polymorphus marilis	*Gammarus lacustris* (fresh-water amphipod)	*Aythya affinis* (Lesser Scaup; duck)	Denny (1968)
P. minutus	*Gammarus* spp.	*Anas platyrhynchos* (Mallard) and other anatid birds	Hynes and Nicholas (1957); Crompton and Harrison (1965)
Pomphorhynchus bulbocolli	*Hyalella azteca*	*Catostomus commersoni* (sucker; fresh-water fish)	Jensen (1952)
P. laevis	*G. pulex*	*Squalius cephalus* (chub)	Ginetsinskaya (1961); Chubb (1965)
Profilicollis botulus	*Carcinus maenas* (crab)	*Somateria mollissima* (Eider duck)	Garden, Rayski and Thom (1964)
Prosthorhynchus formosus	*Armadillidium vulgare* *Porcellio laevis* *P. scaber* (terrestrial isopods)	*Turdus migratorius* (N. American robin and other birds)	Schmidt and Olsen (1964)

TABLE 1. (*cont.*)

Species	Intermediate hosts	Final hosts	References
Eoacanthocephala			
Neoechinorhynchus cylindratus	*Cypria globula* (fresh-water ostracod)	*Huro salmoides* (bass—final host) *Lepomis pallidus* (bluegill—transport host)	Ward (1940*a*)
N. emydis	*Cypria maculata* (1st int. host) *Campeloma rufum* (fresh-water gasteropod; 2nd int. host)	*Graptemys geographica* (Map turtle)	Hopp (1954)
N. rutili	*Cypria turneri*	*Cyprinidae* (and other families of fresh-water fish)	Merritt and Pratt (1964)
Octospinifer macilentis	*Cyclocypris serena* (fresh-water ostracod)	*Catostomus commersoni*	Harms (1965)
Paulisentis fractus	*Tropocyclops prasinus* (fresh-water copepod)	*Semotilus atromaculatus* (Creek chub)	Cable and Dill (1967)

arthropods and must, therefore, acquire cystacanths through eating infected transport hosts.

The preceding discussion and the information summarized in table 2 indicate that, in a life cycle involving no transport host, an egg must be adapted to live in five environments, an acanthor in two or three environments, an acanthella in one environment, a cystacanth in three environments and an adult worm in one environment. As indicated by its title, this book is ecological and, since ecology is the study of the relationship between an organism and its environment, the aim of the book is to relate acanthocephalan physiology to the environments occupied at different stages of the life cycle. The ecological aspect of parasitism was also emphasized by Rogers (1962), who warned that investigations of parasitism are difficult because the relationship between the parasite and the host is in a state of equilibrium which is easily disturbed by experimental intervention. The approach adopted here permits the relationship to be explored, and facilitates identification of the points where the host–parasite equilibrium has been disturbed in the course of investigations and, consequently, where new studies are needed.

TABLE 2. *The environments occupied by the stages of Acanthocephala during the life cycle*

		Stage of the life cycle				
	Environment	Egg	Acanthor	Acanthella	Cystacanth	Adult
1	Region of the final host's intestine where reproduction occurs	+	−	−	+	+
2	Faeces of the final host	+	−	−	−	−
3	Habitat of the final host	+	−	−	−	−
4	Alimentary tract of the intermediate host	+	+	−	−	−
5	Intestinal tissues of the intermediate host	−	+	−	−	−
6	Haemocoele of the intermediate host	−	+	+	+	−
7	Alimentary tract of the transport host	−	−	−	+*	−
8	Abdominal tissues of the transport host	−	−	−	+*	−
9	Alimentary tract of the final host	−	−	−	+	−
10	Pseudocoelom of the female worm	+	−	−	−	−

* Not an obligatory feature in the life cycle of every acanthocephalan.

CHAPTER 2

The Environment of Adult Acanthocephalans in Their Final Hosts

ADULT Acanthocephala achieve sexual maturity in the alimentary tract of a vertebrate. The worms occupy particular locations, some of which are listed in table 3 and illustrated diagrammatically in fig. 4. The fact that the location of so few environments is known when over 650 species of Acanthocephala have been described is disappointing, and lack of attention to this point has probably resulted from the identification of many preserved acanthocephalans by experts who never saw the worms in their hosts. The data discussed below will show that different conditions occur in different parts of the intestine; to state, therefore, that the intestine is the environment of an acanthocephalan is almost as imprecise as stating that the sea is the environment of herrings.

Few conclusions can be drawn from the information in table 3, except that acanthocephalans in birds tend to inhabit the posterior part of the small intestine while those in mammals appear to inhabit the anterior part. Insufficient observations exist about the attachment zones of acanthocephalans from amphibians and reptiles, and the alimentary tracts of fish are so variable (Barrington, 1957) that no pattern of attachment can be expected until the locations of many more acanthocephalans have been reported.

All adult acanthocephalans become attached by their proboscides to their host's intestinal wall. The mechanism of attachment has been investigated by Hammond (1966*b*; 1967*b*) and accounts of the mechanical damage to host intestinal tissues have been given for *Polymorphus minutus* by Crompton (1963) and for a variety of acanthocephalans infecting fish by Chaicharn and Bullock (1967). One effect of attachment is that worms which are small compared with the dimensions of their host's intestine are brought into intimate contact with the

TABLE 3. *The location of the environment of selected acanthocephalan parasites*

	Host	Environment	References
		Fish hosts	
Palaeacanthocephala			
Acanthocephalus jacksoni	*Salvelinus fontinalis* *Salmo gairdneri*	Ileo-rectal regions of intestine	Bullock (1963)
A. lucii	*Perca fluviatilis*	Stomach	Rawson (1952)
Echinorhynchus gadi	*Oncorhynchus kisutch* *O. tschawytscha*	Lower intestine	Ekbaum (1938)
E. lageniformis	*Platichthys stellatus*	Proximal loop of intestine	Ekbaum (1938); Prakash and Adams (1960)
E. truttae	*Salmo trutta*	Whole intestine—sex. mature worms in posterior position	Awachie (1966)
Leptorhynchoides thecatus	*Ambloplites rupestris* *Huro salmoides*	Pyloric caeca	DeGiusti (1949*a*); Venard and Warfel (1953)
Micracanthocephalus hemirhampus	*Hemirhamphus intermedius*	Stomach	Baylis (1944)
Pomphorhynchus bulbocolli	*Catostomus commersoni*	Between middle and posterior intestine	Chaicharn and Bullock (1967)
Telosentis tenuicornis	*Leiostomus xanthurus*	Hind gut	Huzinga and Haley (1962)
Eoacanthocephala			
Neoechinorhynchus carpiodi	*Carpoides cyprinus*	Anterior small intestine	Dechtiar (1968)
N. cristatus	*Catostomus commersoni*	Posterior part of intestine	Chaicharn and Bullock (1967)
N. prolixoides	*Erimyzon oblongus*	Posterior part of intestine	Chaicharn and Bullock (1967)
Neoechinorhynchus spp.	*Coregonus hoyi*	Middle of intestine	Cross (1934)
Octospinifer macilentis	*Catostomus commersoni*	Posterior part of intestine	Chaicharn and Bullock (1967)
O. torosus	*C. occidentalis*	Middle third of intestine	Van Cleave and Haderlie (1950)
Octospiniferoides chandleri	*Gambusia affinis*	Posterior part of intestine	Bullock (1967)
Pallisentis nagpurensis	*Ophiocephalus striatus*	Duodenum and intestine	Bhalerao (1931)
Paulisentis fractus	*Semotilus atromaculatus*	First flexus of intestine below stomach	Cable and Dill (1967)

TABLE 3 (*cont.*)

	Host	Environment	References
Palaeacanthocephala		Amphibian hosts	
Acanthocephalus ranae	1 *Bufo bufo*	Anterior part of small intestine (fig. 4*c*)	D. W. T. Crompton (unpubl. observ.)
	2 *Rana temporaria*	Upper part of small intestine	Pflugfelder (1949)
		Reptilian hosts	
Sphaerechinorhynchus rotundocapitatus	*Pseudechis porphyriacus*	Lower part of intestine and rectum	Johnston and Deland (1929)
Eoacanthocephala			
Neoechinorhynchus emydis	*Graptemys geographica*	Attached to duodenal wall between entry of bile duct and tail of adjacent pancreas	Dunagan (1962)
Archiacanthocephala		Avian hosts	
Mediorhynchus grandis	*Quiscalus quiscala*	Lower portion of small intestine	Moore (1962)
Palaeacanthocephala			
Apororhynchus amphistomi	*Wilsonia canadensis*	Cloacal region	Byrd and Denton (1949)
Arhythmorhynchus capellae	*Capella gallinago delicata*	Small intestine and caeca	Schmidt (1963)
Corynosoma bipapillum	*Larus philadelphia*	Lower intestine	Schmidt (1965)
Polymorphus minutus	*Anas platyrhynchos*	Posterior part of intestine (fig. 4*a*, *b*)	Crompton and Whitfield (1968*b*)
Profilicollis botulus	*Somateria mollissima*	Posterior small intestine	Threlfall (1968)
Prosthorhynchus formosus	*Turdus migratorius*	Posterior part of intestine	Schmidt and Olsen (1964)
Archiacanthocephala		Mammalian hosts	
Macracanthorhynchus hirudinaceus	Domestic pigs	1 Duodenum	Schwartz (1929)
		2 Jejunum	Kates (1944)
Moniliformis dubius	Laboratory rats	Anterior small intestine (fig. 4*e*)	Holmes (1961)
Onicicola canis	Dogs	Jejunum and ileum	Van Cleave (1920*b*)
Palaeacanthocephala			
Corynosoma strumosum	1 *Phoca richardii*	Anterior small intestine	Ball (1928)
	2 *Erignathus barbatus*	Small intestine	Lyster (1940)
Polymorphus paradoxus	*Castor canadensis*	Jejunum	Connell and Corner (1957)
Bolbosoma balaenae	*Balaenoptera acutorostrata*	Duodenum	Barker and Macalister (1865)

intestinal tissues in the region sometimes called the paramucosal lumen (Read, 1950). Large worms will also have some contact with their host's intestinal tissues, but a greater proportion of their surface will be exposed to the lumen.

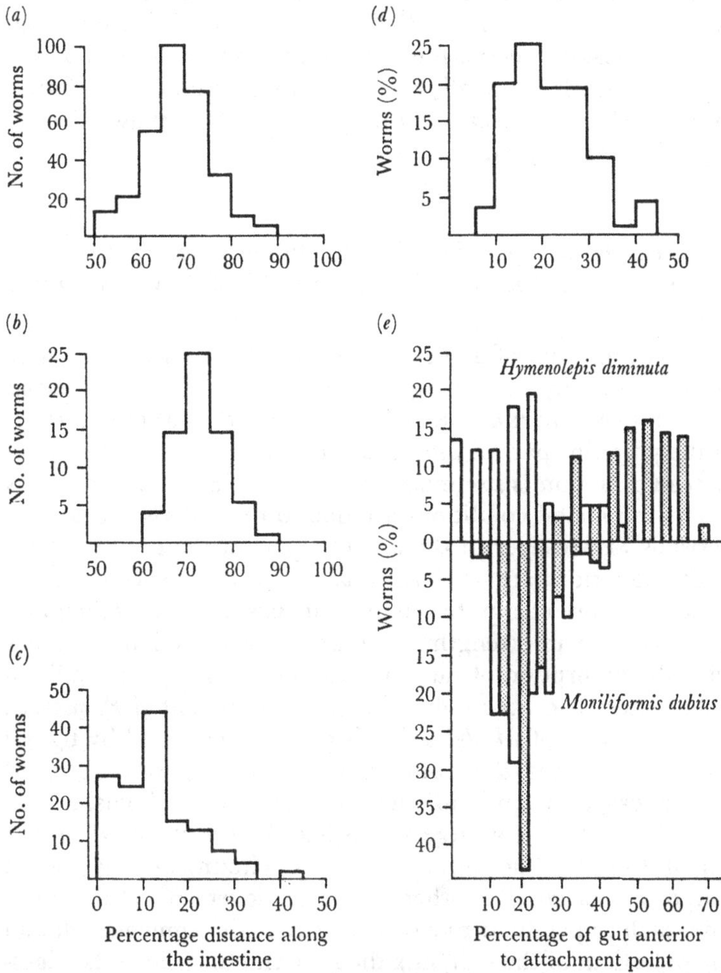

Fig. 4. Diagrammatic representations of the attachment positions of acanthocephalans in their final hosts. (*a*) 1- to 5-day-old *Polymorphus minutus* in domestic ducks (after Crompton and Whitfield, 1968*b*, fig. 3); (*b*) 31- to 35-day-old *P. minutus* in domestic ducks (after Crompton and Whitfield, 1968*b*, fig. 3); (*c*) *Acanthocephalus ranae* in the toad, *Bufo bufo*; (*d*) *Moniliformis dubius* in rats (constructed from Holmes, 1961, fig. 3); (*e*) *Moniliformis dubius* and *Hymenolepis diminuta* in rats (after Holmes, 1961, fig. 1). □, single; ▦, concurrent.

For convenience, the components of an environment may be classified as physical and biotic factors, but it should be remembered that every environment is holocoenotic and all the factors are interdependent (Allee *et al.* 1949). It is impossible here to consider all the literature on the intestines of homeothermic animals, especially mammals, and the reader is referred to the books of Alvarez (1939), Spencer (1960), Wilson (1962) and Wiseman (1964).

The environments of Moniliformis dubius *and* Polymorphus minutus *as examples of the environments provided by homeothermic hosts*

The complexity of any acanthocephalan's environment may be illustrated by describing and comparing that of *M. dubius* in the anterior part of the rat's intestine with that of *P. minutus* in the posterior part of the duck's intestine (table 3). The relevant information is summarized in table 4. Several differences in environmental conditions are due to the different locations. Peristalsis is stronger in the anterior part of the intestine than in the posterior part, so that *M. dubius* may be expected to use relatively more energy to maintain its position than *P. minutus*. The decrease in strength of peristaltic contractions together with the absorption of nutrients causes roughage and indigestible material to accumulate in the environment of *P. minutus*, but not in that of *M. dubius*. This point is illustrated by fig. 5*a* and it is assumed that, although these results were obtained from ducks, they apply to other homeothermic animals. Movement for a relatively large worm like *M. dubius* would be impeded if material accumulated in its environment as it does in that of *P. minutus*. The other muscular movements of the intestine, such as pendular movements, rhythmic segmentation and the pulsations of the villi, mix the nutrients and the hosts' digestive enzymes and disperse the parasites' excretory products.

The fact that the environment of *M. dubius* is drier than that of *P. minutus* (table 4) suggests that *M. dubius* may be adapted to withstand water shortage. Read and Rothman (1958), who were studying the effects of carbohydrate deprivation of rats on *M. dubius*, recorded enormous changes in

TABLE 4. *An analysis of the environments of* Polymorphus minutus *and* Moniliformis dubius *in their final hosts*

		Polymorphus minutus (in ducks)	*Moniliformis dubius* (in rats)
Location/attachment zone		60–80% of distance along intestinal length (Crompton and Whitfield, 1968*b*)	5–35% of distance along intestinal length (Holmes, 1961)
Physical factors			
Space	Volume	15 cm³*	4 cm³*
	Attachment area	80 cm²	40 cm²*
	Host absorptive area	16,000 cm² (Crompton, 1969)	24,000 cm²*
Substratum		Pulsating villi	Pulsating villi
Peristalsis		Weak	Strong
Medium		Aquatic, 87% water plus much intestinal debris (Crompton and Nesheim, 1970)	Aquatic, 77% water plus relatively little debris*
Osmotic pressure		Between 170·5 ± 13 and 178·2 ± 5·5 mM NaCl/l (Crompton and Edmonds, 1969)	Summarized by Follansbee (1945)
Oxygen tension (paramucosal lumen)		From 25 mm Hg to < 0·5 mm Hg (Crompton, Shrimpton and Silver, 1965)	From 30·2 mm Hg to 7·9 mm Hg (Rogers, 1949)
Carbon dioxide tension		40 to 60 mm Hg†	40–60 mm Hg†
Hydrogen ion concentration		7·4 (Crompton, 1966)	6·93 to 7·51 (Kofoid, McNeil and Cailleau, 1932)
Temperature		41·7 ± 0·7 °C (Crompton, 1966)	37·4 °C
Carbohydrates		Mean glucose concentration about 5·7 mg/ml when present (Crompton, 1966)	Expected to be variable
Lipids		1·2% of dry weight of intestinal contents when diet contains 4% lipid (Crompton and Nesheim, 1970)	Expected to be similar to lipid concentration in the rat's diet
Amino acids		All common acids available from both exogenous and endogenous sources (Crompton and Nesheim, 1969*b*)	Some data in Arme and Read (1969)
Bile acids		5·18 mg/g intestinal contents; 59% chenodeoxycholic acid, 16% cholic acid, 17% lithocholic acid and 8% other acids (Crompton and Nesheim, 1970)	80% cholic acid (Haselwood, 1964), chenodeoxycholic acid, ursodeoxycholic acid and α- and β-muricholic acid (Smyth and Haselwood, 1963)

TABLE 4. (*cont.*)

	Polymorphus minutus (in ducks)	*Moniliformis dubius* (in rats)
Hosts' digestive enzymes	Little activity	Considerable activity
Biotic factors		
Micro-organisms	*Escherichia coli*; *Clostridium perfringens*; *Bacillus* spp; Enterococci (Hawker *et al.* 1960)	Few bacteria in the duodenum and jejunum, varied 'flora' in ileum (Wilson and Miles, 1964)
Protozoa	*Eimeria saitamae* (Inoue, 1967) *Wenyonella philiplevinei* (Leibovitz, 1969) *Tyzzeria pernisiosa* (Allen, 1936) *Trichomonas anatis* *Cochlosoma* spp. *Hexamita* spp. *Chilomastix* spp. (Becker, 1959)	
Metazoa	*Dicranotaenia coronula*; *Diorchis stefanskii*; *Schistocephalus solidus* (Crompton, 1969) *Fimbraria fasciolaris* (P. J. Whitfield, pers. commun.)	*Hymenolepis diminuta* (Holmes, 1961)

* D. W. T. Crompton, unpubl. observ. † Estimations from literature.

the wet weights of *M. dubius* when their hosts were starved and fed again over short periods (p. 38). These observations suggest that *M. dubius* may sometimes withstand conditions of near dehydration which could be caused by a rise of intestinal osmotic pressure when a rat, deprived of dietary water, absorbs much of its intestinal water. The reverse situation would occur when the hosts drank water which would lower the environmental osmotic pressure below that of the pseudocoelomic fluid of *M. dubius* and initiate the flow of water into the worms. This explanation of some of Read and Rothman's results depends upon acanthocephalans being osmoconformers rather than osmoregulators, an assumption supported by measurements on the osmotic pressure in the environment of *P. minutus* (table 4). The mean osmotic pressures at the anterior and posterior limits of the normal environment of *P. minutus* are 170·5 and 178·2 mM-NaCl/l respectively and the osmotic pressure of the pseudocoelomic fluid of this parasite is about 178 mM-NaCl/l (Crompton and Edmonds, 1969).

The tensions of oxygen and carbon dioxide in the environments of *M. dubius* and *P. minutus* are likely to be similar (table 4). In addition to these gases, nitrogen, hydrogen sulphide and traces of other gases may be present (Spencer, 1960). Gases in the intestinal lumen originate from diffusion from the vascular system of the intestinal wall and also from swallowed air and bacterial activity. A greater bacterial flora may be expected in the environment of *P. minutus* than of *M. dubius* which may not, therefore, be subjected to the same range of gases as *P. minutus*.

Some differences between the organic substances in the two environments are illustrated in fig. 5. These eight histograms are based on the results of Crompton and Nesheim (1969*a*, *b*, 1970), who studied various aspects of digestive physiology in ducks. If the intestines of birds and non-ruminant mammals are basically similar, the data given for sections A and B in the histograms of fig. 5 may be like those in the environment of *M. dubius* in rats. The absorption of different substances in the intestine of the duck and the rat are compared in table 5, where it can be seen that, of the parasites' main nutrients, higher lipid concentrations, but smaller amounts of free amino acids are likely to be present in the environment of *M. dubius* than of *P. minutus*; the amount of carbohydrate will probably be similar in both environments.

The variety and concentration of amino acids in the environment are important factors for acanthocephalans and cestodes, both of which feed by absorbing nutrients through their surfaces. For some time, attention has been drawn to amino acids in the intestinal lumen by experiments indicating that, during digestion in dogs, exogenous protein is diluted by sufficient endogenous protein to produce a relatively constant mixture of amino acids (see Nasset, 1968). These results were obtained by detecting the amino acids present in the intestine $1\frac{1}{2}$ h after dogs, which had been starved for several hours, had been given a test meal of known composition. Pancreatic secretions and, to a lesser extent, gastric secretions, mucus and debris from the turnover of the intestinal epithelium are the main sources of endogenous protein. The immediate significance of Nasset's

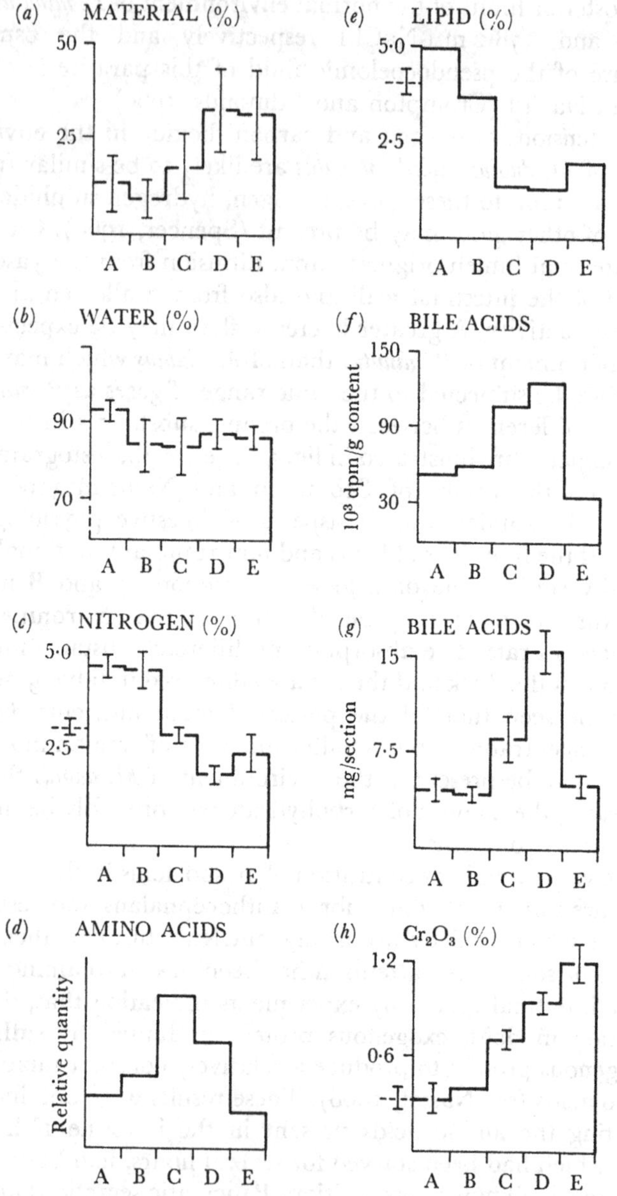

Fig. 5. Histograms showing the distribution of (*a*) % material; (*b*) % water; (*c*) % nitrogen; (*d*) amino acids; (*e*) % lipid; (*f*) and (*g*) bile acids; (*h*) % chromic oxide. (After Crompton and Nesheim, 1969*a*, *b*, 1970). A, B, C, D and E on the abscissae represent a fifth of the distance along the intestine. *Polymorphus minutus* inhabits region D.

TABLE 5. *A comparison of the absorptive functions of the parts of the intestine which form the environments of* P. minutus *and* M. dubius

		Polymorphus	*Moniliformis*
Monosaccharides		++	+++
Disaccharides		+++	++
Amino acids	exogenous	++	+++
	endogenous	++	+
Lipids		+	+++
Cholesterol		+++	o
Bile acids		+++	o
Water		++	+
Vitamin B_{12}		+++	+
Vitamin A		+	+

+++ to o indicates importance of this region of the intestine as a site for absorption. Data based on Wilson (1962); Wiseman (1964); Crompton and Nesheim (1969*b* and 1970).

findings was that the carrier mechanisms for amino acids in the absorptive surface of a parasite may have become so intimately adapted to the constant amino acid mixture in the intestine of a given host that the constancy of this mixture could be responsible for governing host–parasite specificity (Read, Rothman and Simmons, 1963).

Experiments to discover the amino acid pattern in the intestine of ducks have been undertaken by Crompton and Nesheim (1969*a*, *b*), although the ducks were allowed to feed *ad libitum.* Under this condition, the pattern of the anterior and middle parts of the intestine reflects that of the diet, but the pattern is more stable posteriorly in the environment of *P. minutus* (section D, fig. 5*d*), where it is similar to that found elsewhere in the intestine for ducks on a protein-free ration. If most endogenous protein in the intestine originates from the pancreas, it is likely that this protein will not have been exposed to gastric conditions. The component amino acids, therefore, will take longer to be released during digestion and will be absorbed more posteriorly. Consequently, *P. minutus* in ducks and many acanthocephalans in other birds (table 3) may experience stable amino acid patterns, while *M. dubius* and acanthocephalans in the anterior intestine of other mammals may be subjected to changing patterns unless they migrate in the intestine during digestion.

Another difference between the environments of *M. dubius* and *P. minutus* is the nature of the bile acids present (table 4). The decrease in the strength of peristaltic contractions results in an accumulation of bile acids in the environment of *P. minutus* (fig. 5*f*, *g*). This condition is found in many animals; the posterior region of the intestine of rats is the principal site of the reabsorption of bile acids (table 5). Some aspects of the effects of bile acids on parasites are discussed below (p. 48), but it is interesting to note here that *P. minutus* lives in a detergent solution.

There is likely to be more activity from the host's digestive enzymes in the anteriorly situated environment of *M. dubius* than in the posteriorly situated one of *P. minutus*. This can be inferred from the recent work of Zoppi and Shmerling (1969), who demonstrated a greater quantity and variety of enzyme activity in the duodenum and jejunum than in the ileum of turtles, birds and mammals.

Some of the biotic factors in the environments of *M. dubius* and *P. minutus* are listed in table 4. Most text books of bacteriology state that fewer micro-organisms live in the anterior part of the intestine than in the posterior part and the same generalization probably applies to Protozoa inhabiting the small intestine. Thus, the competitors of *M. dubius* are most likely to be individuals of its own species or other helminths—for example, *Hymenolepis diminuta*—which normally occupy this environment. Competiton consists of exploitation and interference (Park, 1962). Exploitation is observed when animals draw upon a nutrient which is present in limited supply while interference occurs when individuals disturb each other's feeding or reproductive behaviour. Holmes (1961) has found that *M. dubius* shows a crowding effect for infections involving 40 cystacanths and also when the cestode, *H. diminuta*, is present concurrently. He suggests that competition for carbohydrates could cause this effect. Since both types of worm absorb their nutrients through their surfaces, exploitation rather than interference is likely to be involved. In concurrent infections of *M. dubius* and *H. diminuta*, the cestodes are usually displaced to a more posterior position in the environment (fig. 4*e*). Ward and Crompton (1969) suggest that the ethanol excreted by *M. dubius* could be responsible for this effect (p. 50) and the movement of the cestodes to a less favourable environment could account for their stunted appearance; but in hamsters Holmes (1962) has found that *M. dubius* and *H. diminuta* share the same environment and show no signs of a crowding effect.

Male *M. dubius* from rats have been reported to be smaller when female worms are present than when they are absent (Graff and Allen, 1963). This effect may be due to competition between the sexes for a limited supply of some nutrient or may be due to the secretion by the female worms of some substance which represses the male worms. These observations of Graff and Allen and those of Holmes illustrate some of the complex interactions which occur between helminths in the intestine. Research of this type has still to be started with *P. minutus* and other acanthocephalans from homeothermic hosts.

The environment of acanthocephalans in fish

In general, fish appear to feed on whatever food is available during a particular season and, consequently, their alimentary tracts tend to be unspecialized (Barrington, 1942, 1957). The significance of this principle for the parasitologist is that, until more research has been done, a particular function cannot be assigned to a particular region of the intestine with the same accuracy as for a bird or mammal. The anatomy of the alimentary tract is also very variable. Many fish do not possess a stomach—for example—*Catostomus catostomus* (Weisel, 1962), others have pyloric caeca, and so on. Several features, however, of the intestines of fish will be essentially the same as described above for rats and ducks. Roughage accumulates posteriorly as a result of stronger peristalsis and nutrient absorption anteriorly. Similarly, the pH tends to remain approximately neutral along the alimentary tract. The studies of Dawes (1929, 1930) and Al-Hussaini (1949*a*, *b*) provide some information about fish intestines in a manner which is useful when considering helminth environments.

Digestion in fish takes longer than in homeothermic animals and is greatly influenced by the fish's environmental temperature. A fish eaten by a pike, *Esox lucius*, may remain in the stomach for 5 days (Barrington, 1942). The pH at the surface of the prey will be about 2·4 to 3·6 while a higher pH will occur in the tissues of the prey where auto-digestion is occurring. These physiological conditions may be of great significance if the pike's prey is a transport host for an acanthocephalan parasite (chapter 9, p. 103). Acanthocephalan parasites of physostomatous fish, for example, *Esox*, *Tinca* and *Salmo*, may be exposed to higher oxygen tensions than many other acanthocephalans, since these hosts swallow air to maintain the gas volume of their swim bladders. In some fish, for example, certain members of the family Loricariidae, parts of the intestinal tract are modified as a respiratory organ. Acanthocephalans have been reported from fish of this family and any from hosts with a respiratory intestine would be in contact with higher oxygen tensions than are found in other vertebrate intestines. The oxygen tension in turn would affect the bacterial flora and other features of the intestine.

There are bound to be fluctuating environmental conditions for acanthocephalans infecting salmon, eels or other fish which undergo spawning migrations or other movements between fresh and salt water. While young salmon are living in fresh water during their first 3 or 4 years of life, *Neoechinorhynchus rutili* is a common parasite, but it is lost and replaced by *Echinorhynchus gadi* once the fish reach the sea (Dogiel, 1961). Conversely, acanthocephalans acquired during the salmon's marine phase are lost when the fish return to spawn. An investigation of the changes in the salmon's intestinal physiology would probably give much information about the environmental requirements of *N. rutili* and *E. gadi*.

Interactions of the type discussed above for *M. dubius* in rats (p. 22) are to be expected for acanthocephalans in fish. A particularly interesting example of an interaction between an acanthocephalan, *Neoechinorhynchus* spp., and a cestode, *Protocephalus exigus*, has been described by Cross (1934) from ciscoes, *Coregonus hoyi*. Hosts infected with 15 or more acanthocephalans rarely contain more than 4 cestodes while those with 25 or more cestodes usually have few or no acanthocephalans. The cestode inhabits the pyloric caeca of *C. hoyi* and the acanthocephalan lives in the middle portion of the host's intestine (table 3). Direct competition, therefore, of the type involving *M. dubius* and *H. diminuta* is unlikely to provide the explanation for Cross's observations. He suggests that a non-specific immunity limits either the cestodes or the acanthocephalans when one of these parasites is present in large numbers. Recently a hypothesis of non-reciprocal cross-immunity, which could account for these observations in ciscoes, has been proposed by Schad

(1966). The antigenic stimulus of one helminth is considered to elicit an antibody response by the host, but these antibodies are effective against other helminths which, on infecting that host, find it to have been immunized against them. Thus, the first arrivals from *Neoechinorhynchus* or *Protocephalus* in the intestine of *C. hoyi* stimulate an antibody production effective against the second arrival. In fact, these are hypothetical considerations and no detailed studies of the immunological responses of hosts to acanthocephalans appear to have been undertaken.

That discussion of fish intestines is justified because many acanthocephalans have been described from fish. The environment, however, of any of these worms cannot be described as thoroughly as those from homeothermic hosts. There is no reason to discuss the intestinal conditions of amphibians and reptiles considering the relatively small number of their acanthocephalans that are known, but one important point must be mentioned. Many amphibians and reptiles fast and hibernate during the year and yet do not lose their acanthocephalan parasites. The intestines of turtles, which have stopped feeding, are known to remain suitable environments for *N. emydis* (Van Cleave and Ross, 1944) and frogs, which have not fed for 4 months, can still contain *Acanthocephalus ranae* (Pflugfelder, 1949); acanthocephalans from homeotherms usually die in a few days if their hosts are starved.

Since different conditions prevail in different regions of the vertebrate intestinal tract, some Acanthocephala which are long compared with their host's intestine may be subjected to different conditions at their anterior and posterior ends. That must be true of *M. clarki* found by Benton (1954) in pine mice, for some of the worms extend for virtually the entire intestinal length. Podder (1937) found a sexually mature *N. topseyi* which extended from its attachment point in the duodenum to the cloaca of a mango fish. These worms and, to a lesser degree, those like *M. dubius* in rats, are subjected to different environmental conditions along their length at the same time.

MOVEMENTS OF ACANTHOCEPHALANS WITHIN THEIR ENVIRONMENTS

All acanthocephalans must move to some extent within their environments in order to mate, and the evidence suggests that males move more than females. It is not known whether an acanthocephalan returns to its original point of attachment after mating.

If acanthocephalan parasites actually change their attachment zones in their host's intestines, they may also change their environments. During the course of infections of *P. minutus* in ducks, young worms have a mean point of attachment at about

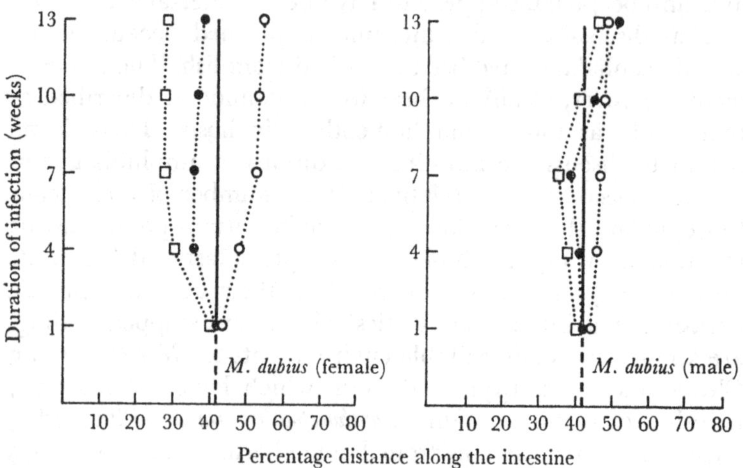

Fig. 6. Graphs indicating the positions of female and male *Moniliformis dubius* during the course of the infection in rats. Ordinate represents duration of infection (weeks). Abscissa represents percentage distance along the intestine. ●···●, Mean positions of centres of worms of average length; ○···○, mean positions of centres of worms of average length assuming no migration; □···□, mean attachment positions; ———, initial mean attachment position. (After Crompton and Whitfield, 1968*a*, figs. 2 and 3).

70 % of the intestinal length while older worms are found at 74 % of the distance along the intestine (fig. 4*a*, *b*). This posterior migration of about 6 cm for a population of worms will not result in an environmental change other than relief from any osmotic stress imposed by the greater fluctuations in osmotic pressure at the anterior part of their environment (Crompton and Edmonds, 1969).

On the other hand, *M. dubius*, the females of which alter in length from a few mm to 20 cm, must be in danger of protruding beyond its normal attachment zone unless anterior migration occurs. This parasite has been observed to become established initially in the middle region of the rat's intestine and then to move forward (Burlingame and Chandler, 1941; Holmes, 1961) and calculations from these authors' data confirm that the migration prevents *M. dubius* from experiencing a major change of environment (fig. 6).

During the course of infection of *E. truttae* in brown trout, *Salmo trutta*, Awachie (1966) found that the worms migrate from the pyloric caeca to the posterior intestine, which is the only site where mature females are found. This migration of *E. truttae* takes about 10 weeks to complete and must involve several changes of environment. Similarly, immature individuals of *Acanthocephalus jacksoni* are always positioned more anteriorly in the the intestines of trout than mature specimens, an observation indicating that this species also undergoes a considerable posterior migration (Bullock, 1963).

An attempt has been made above to define and analyse the environments of a few adult acanthocephalan worms. Sufficient information has been collected about these environments to expose the complex nature of the host–parasite relationship and the need for many more investigations.

CHAPTER 3

Feeding of Adult Worms

THE nutrients required by an adult acanthocephalan for growth and development are obtained from the contents of that part of the host's intestine in which the worm lives. Convincing evidence for this statement has been obtained by Edmonds (1965), who gave oral doses of ^{32}P disodium hydrogen phosphate to some rats infected with *Moniliformis dubius* and intraperitoneal doses to others. The recovery of isotope in worms infecting rats which had received the oral dose was far greater than that from rats which had been injected intraperitoneally; much isotope, however, was recovered from the intestinal mucosa of these rats. Similar results were obtained when Edmonds used ^{14}C L-leucine. This demonstration has largely nullified the implication of Van Cleave (1952) that the attachment organs (praesoma) could have a secondary function of facilitating absorption from the host's tissues by ensuring close contact between the worm's surface and the intestinal villi. Furthermore, circumstantial evidence in favour of absorption from the intestinal contents is provided by the fact that the metasoma, which forms most of the surface, is exposed in the lumen. In the case of *Polymorphus minutus*, the absorptive surface of the praesoma, which is always in contact with host tissues, comprises about 3% only of the total surface of the worm (Crompton, 1969). This finding refutes the suggestion that worms such as *Apororhynchus amphistomi*, which inhabits its host's cloaca (table 3), would primarily obtain nutrients from the tissues (Byrd and Denton, 1949). Finally, the praesoma of both *Acanthocephalus ranae* (Hammond, 1968*b*) and *Paulisentis fractus* (Hibbard and Cable, 1968) appears to play no part in nutrient absorption *in vitro* when studied with autoradiographic techniques.

A list of nutrients, some of which are known to occur in the worms' environments (chapter 2), and other substances which adult acanthocephalans are capable of absorbing is given in table 6. Direct evidence for the uptake of some of these substances is lacking, but the list indicates the types of molecules which can be absorbed.

TABLE 6. *Substances reported to be absorbed by acanthocephalans*

	*M.h.**	*M.d.*	*A.r.*	*P.m.*	*N.* spp.	*P.f.*
D-Glucose	.	+[5,8]	+[6]	+[1]	+[2,3]	+[7]
D-Galactose	.	+[8]	.	.	+[2]	.
D-Mannose	.	+[8]	.	.	.	.
D-Fructose	.	+[8]	.	.	.	.
Maltose	.	+[8]	.	.	+[2]	.
Trehalose	.	.	.	.	+[2]	.
L-Alanine	+[10]	+[10]	.	.	.	.
L-Isoleucine	+[10]	+[10]	.	.	.	.
L-Leucine	+[10]	+[4,10]	.	.	.	.
DL-Methionine	.	+[4]	.	.	.	.
L-Methionine	+[10]	+[10]	+[6]	.	.	.
DL-serine	.	+[4]	.	.	.	.
L-Serine	+[10]	+[10]	.	.	.	.
L-Tyrosine	.	.	.	.	.	+[7]
DL-valine	.	+[4]	.	.	.	.
Glyceryl trioleate	.	.	+[6]	.	.	.
Thymidine	.	.	.	.	.	+[7]
Disodium hydrogen phosphate	.	+[4]	.	.	.	.
Lipid crimson	.	.	+[6]	.	.	.
Sudan IV (Scharlach R)	.	.	+[6,9]	.	.	.
Sudan Black	.	.	+[6]	.	.	.

1 Crompton and Lockwood (1968)
[2] Dunagan (1962)
[3] Dunagan (1964)
4 Edmonds (1965)
5 Graff (1964)
6 Hammond (1968*b*)
7 Hibbard and Cable (1968)
[8] Laurie (1957)
9 Pflugfelder (1949)
10 Rothman and Fisher (1964)
[ ] Indirect evidence.

* *M.h., Macracanthorhynchus hirudinaceus*; *M.d., Moniliformis dubius*; *A.r., Acanthocephalus ranae*; *P.m., Polymorphus minutus*; *N.* spp., *Neoechinorhynchus* spp.; *P.f., Paulisentis fractus.*

NATURE OF THE ABSORPTIVE SURFACE

The electron micrographs of the body wall of *Pomphorhynchus laevis*, *P. minutus*, *A. ranae* and *M. dubius* (p. 30) have provided evidence from which absorption could be adduced as an obvious function of the surface layers. These studies show that the worms' absorptive surfaces have been greatly increased by the development of membrane-lined pores, ducts and canals; and it has been assumed that nutrients pass through the pores and down the canals of the striped layer, and so into the body wall. This assumption, however, lacks precision and produces misconceptions about the absorptive surface and process. For example, Crompton and Lee (1965) and Hammond (1967*a*), who interpreted their studies on *P. minutus* and *A. ranae* respectively by diagrams, depict the canals and ducts of the striped layer disintegrating into vesicles in a manner suggesting pinocytosis, but there is no actual evidence of this. Carbon and thorium dioxide have been detected, with the electron microscope, in the surface canals of *M. dubius* which had been incubated *in vitro* in suspensions of these materials (Edmonds and Dixon, 1966). These authors conclude that their micrographs indicate the uptake of carbon and thorium dioxide but, in fact, substances which have not passed through the plasma membrane covering the cuticle and lining the pores, canals and ducts of the surface layers (fig. 2*d*, *e*; plate 1 B) cannot be said to have been absorbed. This result obtained with electron dense substances is important because it demonstrates that the pores in the cuticle are not an artifact; the pores have since been seen with the stereo-scanning electron microscope in *A. ranae* (Hammond, 1968*a*) and *P. minutus* (plate 2 B). When the extent of the surface's plasma membrane has been determined, an important contribution to our understanding of absorption by Acanthocephala will have been made.

A little more information has recently been obtained about the ultrastructure of the surface of the metasoma of *P. minutus*. By examining tangential sections of the surface, each pore, and the canal into which it leads, is seen to be lined by plasma membrane surrounded by a granular layer which in turn is enveloped by a clear layer (plates 1 A, B). The canal from each pore bifurcates and one of the resultant canals divides again and so on; the deeper the section is cut into the body wall, the more divisions are detected. The evidence for this interpretation is shown in plate 1 B and the arrangement of the granular and clear layers, once the divisions have been reached, is depicted in fig. 7. The effect of the divisions of the canals will be a huge increase in the surface area of the plasma membrane.

Several reports have been made of phosphatase activity in the surface layers of acanthocephalans, the granular and clear layers of the pore-canal units being the obvious sites. Ultrastructural detection of phosphatase activity in the surface canals of *M. dubius* has been made by Rothman (1967) (plate 2A) and

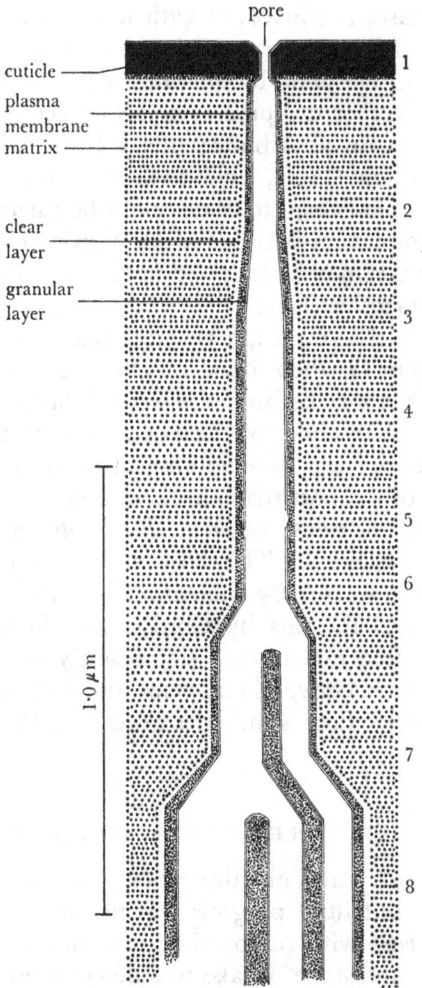

Fig. 7. Diagrammatic reconstruction of the surface region of *Polymorphus minutus*, based on plate 1 B.

this can be predicted to be the location of the alkaline and acid phosphatase activity found in several archi- and palaeacanthocephalans, but not in eoacanthocephalans, by Bullock (1949*a*,

1958, 1960). Phosphatases bring about both the transfer and hydrolysis of a phosphate group for which they are thought to be specific irrespective of the remainder of the molecule. Most living cells do not readily lose phosphate groups and the notion that phosphatases are involved with nutrient absorption may have originated from the assumption that once a substance has been absorbed, phosphorylation will prevent its escape. There is no evidence yet for this process in Acanthocephala and, since eoacanthocephalans absorb nutrients (table 6; Dunagan, 1962, 1964; Hibbard and Cable, 1968) in spite of their apparent lack of surface phosphatases, attention could be turned away from the possible connexion between absorption and these enzymes. The phosphatases may be involved in the growth and maintenance of the worms' surfaces.

It is unlikely that even the heaviest acanthocephalan infections ever deprive their hosts of glucose, amino acids and nutrients of that sort because the worms' surface areas are so much smaller than those of the hosts. About 560 3-week-old *P. minutus* of both sexes have an absorptive surface area of only 1·7 % of that of the host tissue in their environment. Competition for many nutrients between *P. minutus* and ducks will therefore be negligible. The host may compensate for the presence of the worms by increasing its surface area in the manner reported for rats by Gordon and Bruckner-Kardoss (1961). Germ-free rats have a significantly smaller intestinal surface area than conventional ones and the authors suggest that the intestinal flora stimulates growth of the microvilli in these animals.

NATURE OF THE ABSORPTIVE PROCESS

The process of nutrient absorption has been studied in various ways, but little progress was made until worms were incubated *in vitro* with radioactive nutrients. In some investigations the quantitative uptake was determined and in others qualitative observations were made with autoradiographic techniques (table 6).

Carbohydrates and amino acids

The uptake of ^{14}C glucose by *P. minutus in vitro* has been investigated by Crompton and Lockwood (1968) under con-

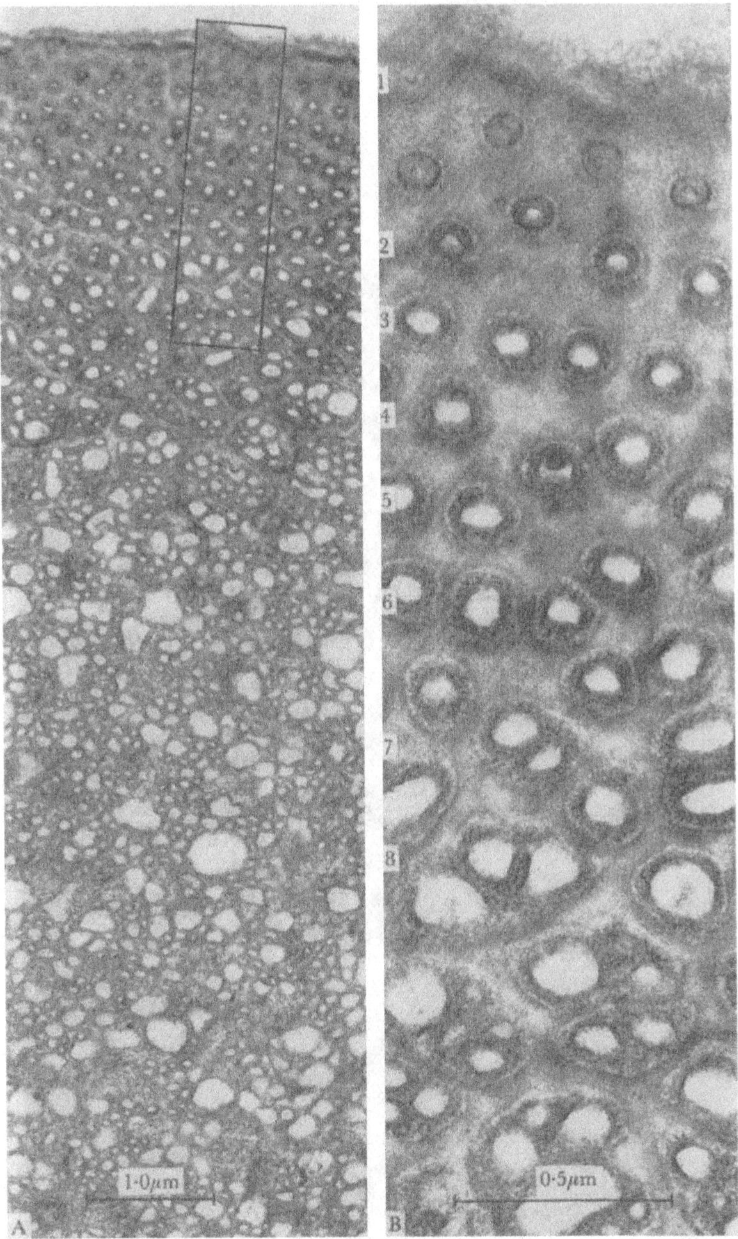

Plate 1. Electron micrographs of tangential sections of the metasomal body wall of *Polymorphus minutus* fixed in glutaraldehyde. A, low magnification; B, higher magnification of the inset on plate 1A.

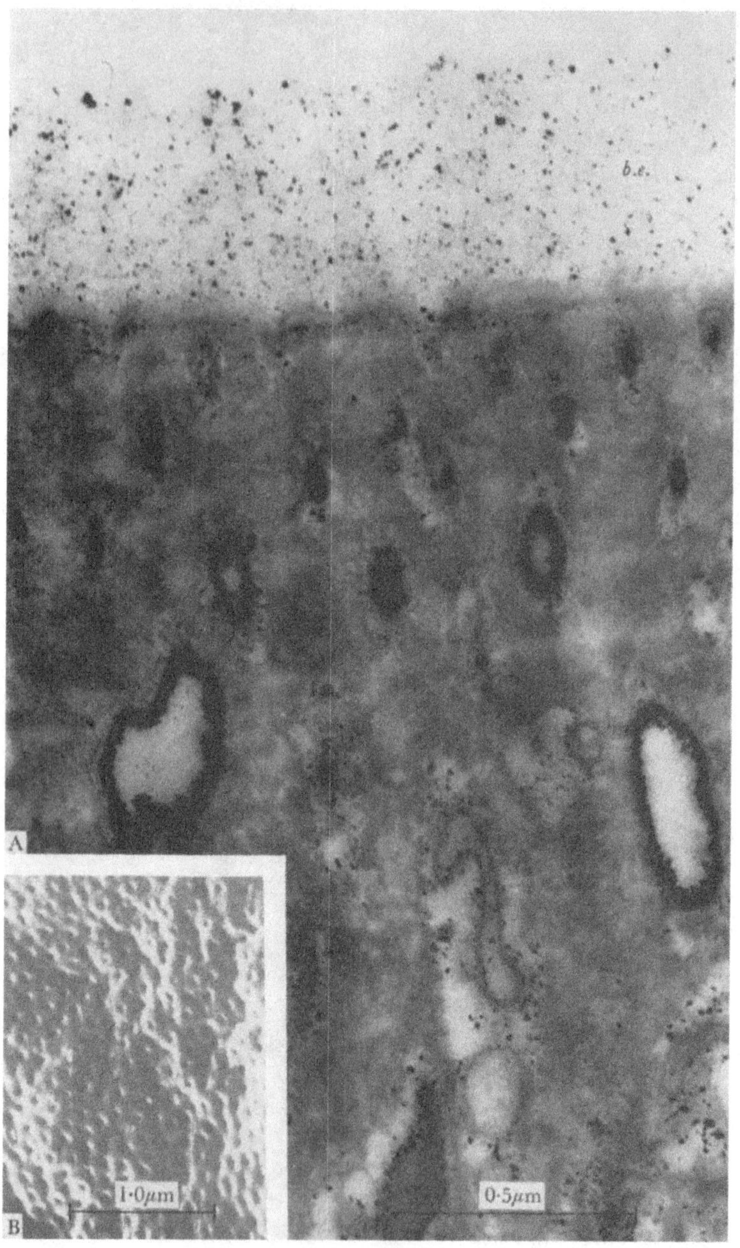

Plate 2. Electron micrographs of the acanthocephalan body wall. A, phosphatase activity associated with the canals near the surface of *Moniliformis dubius*: note the appearance of the epicuticle (after Rothman, 1967; fig. 6); *b.e.*, epicuticle; B, the surface of *Polymorphus minutus* seen with the stereo-scanning microscope (by courtesy of Cambridge Instrument Company).

ditions designed to simulate those in the worm's environment (table 4). The results should have genuine significance because worms could be replaced, by surgical methods, into their environment in uninfected ducks after an incubation lasting 4 h; the worms continued their development and thus were shown to be healthy at the end of the experiment. The results indicate that glucose can be absorbed against a concentration gradient and that its uptake reaches a maximum value of about 12 μg/h

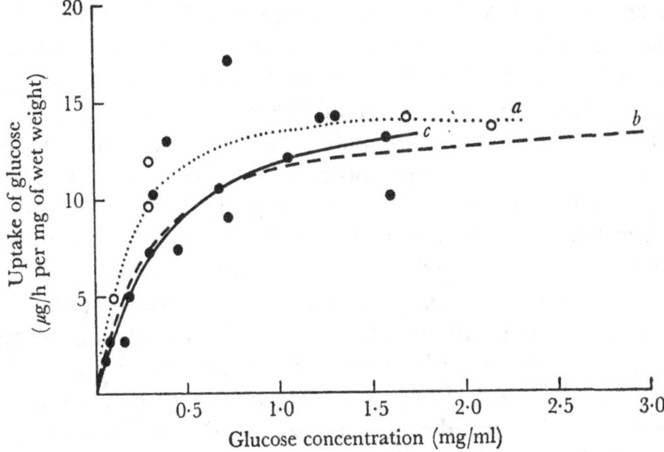

Fig. 8. Graph showing the result of plotting the uptake of glucose by *Polymorphus minutus* against the concentration of glucose in the incubation medium. Open circles represent preliminary results. (After Crompton and Lockwood, 1968, fig. 2).

per milligram of worm once the environmental concentration is about 2 mg/ml (fig. 8). Glucose is absorbed at about half this rate when the environmental concentration is 0·3 mg/ml. An analysis of the data indicates that a carrier system, situated in the surface of *P. minutus*, must be operating and that all the sites are saturated when the concentration is 2 mg/ml.

That *P. minutus* obtains glucose *in vivo* by the same process as that demonstrated *in vitro* need not be doubted, although an environmental glucose concentration of 2·0 mg/ml may be higher than is necessary to saturate the carrier system *in vivo*. Worms acclimatized to a given glucose concentration for an hour *in vitro*, have a maximum uptake rate of only 4–5 μg glucose/h per milligram of worm. Worms living in the intestine are likely to become acclimatized to a given glucose concentration, and if they are attached to their host's intestine their proboscides will be inactive and not requiring energy for the movements so often observed *in vitro*. These considerations suggest that the maximum rate of glucose absorption measured for Acanthocephala *in vitro* may not reflect accurately the rate *in vivo*.

Studies on the uptake of several neutral amino acids by *M. dubius* and *Macracanthorhynchus hirudinaceus* have indicated that acanthocephalans employ active processes to obtain these nutrients (Rothman and Fisher, 1964). L-Methionine is absorbed against a concentration gradient by both sexes of both species of worm and free methionine can be detected in the body fluid of worms within 1 min of the start of an experiment. L-Alanine, L-isoleucine, L-leucine and L-serine are also absorbed in a similar manner and all the amino acids used cause competitive inhibition in *M. dubius* (fig. 9).

Several conditions must be fulfilled before the active uptake of a non-electrolytic nutrient can be demonstrated convincingly (Wilson, 1962). Of these, the crucial ones are movement against a concentration gradient, saturation effects and competitive inhibition. Thus the active uptake of amino acids by *M. dubius* and *M. hirudinaceus* has been established and strong evidence in favour of active uptake of glucose by *P. minutus* has been obtained. Competitive inhibition of glucose uptake by another monosaccharide was not studied for *P. minutus* by Crompton and Lockwood, but galactose is a likely competitor (table 6).

Lipids

So far qualitative methods have been employed to study the absorption of lipids by Acanthocephala, but understanding of the process has been obscured because lipid-soluble dyes rather than actual lipids have been used (table 6). For example, Pflugfelder (1949) forcibly fed pork fat stained with Sudan IV to frogs, *Rana temporaria*, infected with *A. ranae*. Twelve hours later, he collected the worms and noticed that their necks and proboscides were stained and later on that lipid droplets inside the lemnisci were stained. Pflugfelder's work has been repeated recently by Hammond (1968*b*), who fed infected toads, *Bufo bufo*, with three dyes, Sudan IV, Sudan Black and Lipid Crimson, in saturated solution in olive oil. He too recovered *A. ranae* in which the praesoma and lemnisci were stained. In

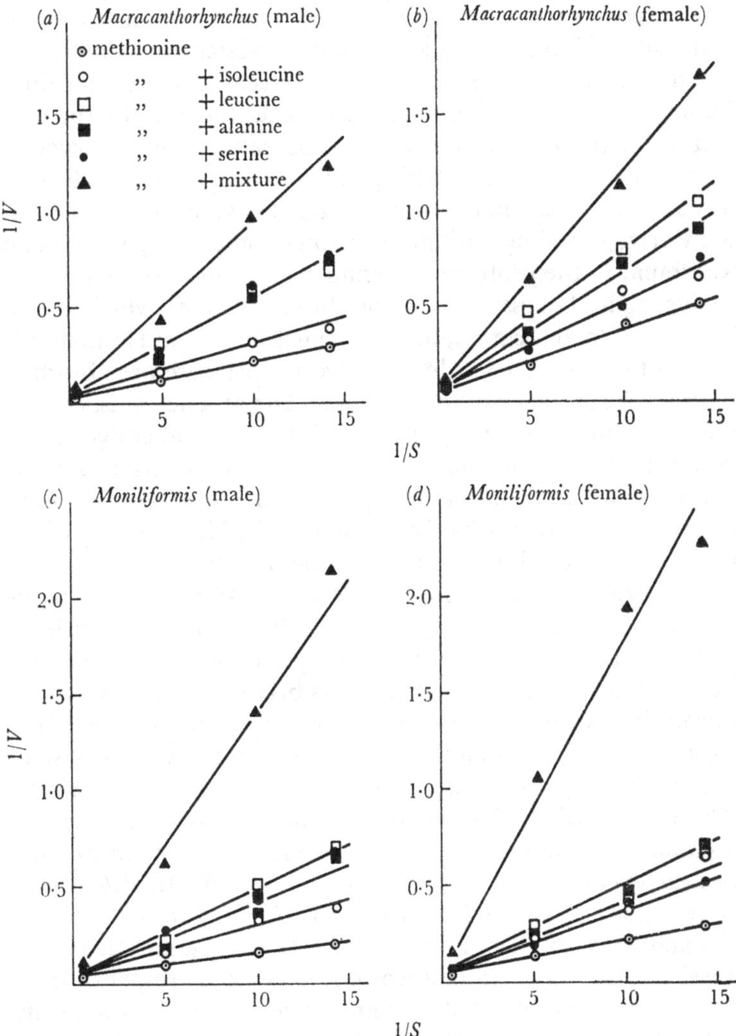

Fig. 9. The uptake of amino acids by *Macracanthorhynchus hirudinaceus* and *Moniliformis dubius*; (*a*), (*b*), (*c*) and (*d*) show double reciprocal plots of L-methionine, L-methionine with various inhibitors, and L-methionine with a mixture of the inhibitors. $1/V$ expressed in μmoles/g h and $1/S$ in mM^{-1}. The curves have been drawn by inspection. (After Rothman and Fisher, 1964, fig. 1).

spite of the fact that the praesoma is known, from histochemical evidence, to be involved in lipid metabolism (Bullock, 1949*a*), these experiments have not necessarily shown that the praesoma is the site of lipid absorption as was suggested by Pflugfelder; the host's digestive processes could have separated the lipid from the dye, facilitating the absorption of the lipid by the host and the dye by the parasite. In the same paper, however, Hammond describes incubating *A. ranae in vitro* with ^{3}H glyceryl trioleate and toad bile. Autoradiographs were prepared from the worms revealing tritium in the metasomal body wall; none was found in the proboscis or lemnisci even after an incubation lasting $1\frac{1}{2}$ h. Unless *A. ranae* produces enzymes which digest lipids prior to absorption, the clear conclusion from Hammond's work is that lipid molecules like glyceryl trioleate are absorbed through the surface of the metasoma. The absence of isotope in the praesoma, but the presence of dyes, is indicative of an excretory function for this group of organs. Circumstantial evidence from electron micrographs for the excretion of lipid through the praesoma has been obtained by Hammond (1967*a*, 1968*a*) and the subject should now be investigated.

Other references to the absorption of substances by acanthocephalans are cited in table 6, but that work adds little to present knowledge about the absorptive process. Some difference in absorption, however, exists between male and female worms. The evidence shows that males contain and, therefore, must have absorbed more glucose per unit weight of tissue than females in the case of *M. dubius* (Graff, 1964) and *P. minutus* (Crompton and Lockwood, 1968). *Moniliformis dubius* has also been found to contain more lipid per unit weight than *M. hirudinaceus* (Beames and Fisher, 1964). Although *M. dubius* lives more anteriorly in the intestine than *M. hirudinaceus*, and may be exposed to higher lipid concentrations (chapter 2, p. 20), the size rather than sex of a worm is likely to be responsible for the difference in rate of absorption, since the smaller the worm, the greater its surface area to volume ratio. For example, young *P. minutus*, which have a surface area to volume ratio twice that of old males and three times that of old females, absorb twice and three times as much glucose per unit weight respectively (Crompton and Lockwood, 1968).

SUMMARY

Acanthocephalans feeding in their hosts obtain their nutrients through the metasomal surface from material in the intestinal lumen of their environments. The body wall is adapted to facilitate absorption because of its increased surface area, which results from the lining of the pores and canals of the surface layers with plasma membrane. The combined absorptive surface in a large acanthocephalan infection is small compared with that of the host's intestine so that competition for many nutrients will be negligible.

The evidence shows that amino acids and monosaccharides are absorbed by active processes. Further investigations of the absorption of these substances need to be undertaken with worms kept in mixtures of amino acids like those found *in vivo* (table 4).

CHAPTER 4

The Influence of the Environment on the Growth and Metabolism of Adult Worms

THE growth of an organism is observed as an increase in its mass resulting from the uptake and utilization of nutrients. Changes in the wet weights of worms may be used to estimate growth provided that the worms are weighed quickly from normal hosts with access to adequate food and water. Evidence to be discussed below will show how temporary interference with the host's diet can influence the wet weight of an acanthocephalan. The growth of *Moniliformis dubius* in rats has been studied by measuring wet weights at regular intervals by Graff and Allen (1963) and their findings are shown in fig. 10. Both male and female worms are bigger and more numerous in male rats than in females. Since all their rats experienced the same feeding routine, Graff and Allen concluded that their results were probably connected with the hormones of female rats. They also observed that male worms grow bigger and synthesize more glycogen when males only are present in an infection than in mixed infections. Thus female worms must retard growth and glycogenesis in males. The implications of these results were not considered when the growth of *Polymorphus minutus* was studied by measuring wet weights (fig. 11). *Polymorphus minutus* has a relatively short life span and thus readily demonstrates that males and females grow at the same rate until copulation, but thereafter, the female's growth rate exceeds that of the males; this growth pattern applies to *M. dubius* (fig. 10) and *Macracanthorhynchus hirudinaceus* (Kates, 1944) and probably to all other species.

The danger of relying on wet weight as a measure of growth, unless the host's dietary requirements are controlled, is illustrated by some work of Read and Rothman (1958), who found that the wet weights of female *M. dubius* changed from 190 mg

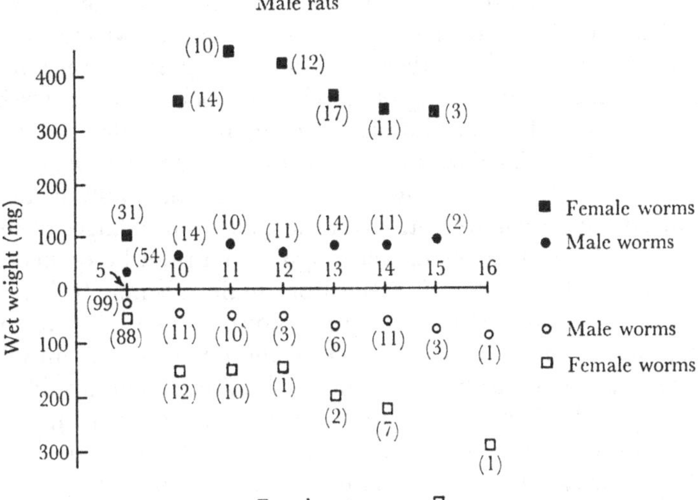

Fig. 10. The development of *Moniliformis dubius* in rats. Numbers in parentheses are the number of worms recovered from three rats. Points on the graph represent the mean weight of worms recovered. (After Graff and Allen, 1963, fig. 1.)

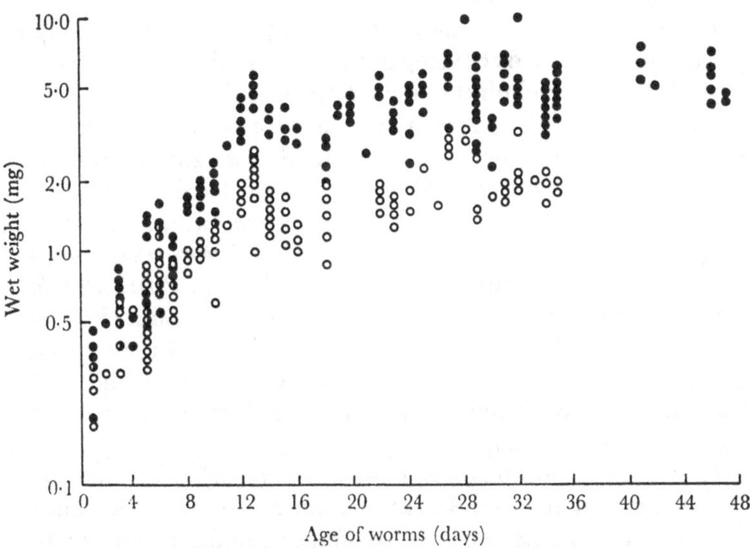

Fig. 11. Semi-log plots of the wet weight of *Polymorphus minutus* against age. Open circles represent male worms and closed circles represent female worms. (Redrawn from data in Crompton and Whitfield, 1968*b*, figs. 5 and 6).

to 84 mg when rats were starved for 48 h. With some older female worms in rats starved for 36 h, the wet weights were found to increase from about 219 mg to 523 mg within 7 h of the host receiving 700 mg of starch and 2000 mg of water by stomach tube. The rat's intestine is comparatively dry (table 4) and although *M. dubius* must be tolerant of considerable dehydration (p. 18), it is curious that the worms in question more than doubled their weight. Until the causes for changes of this sort have been discovered, wet weight should not be trusted as a measure of growth; *M. dubius* from male rats may be bigger than those from females (fig. 10) because female rats eat and drink less than males. Estimations of the nitrogen content of a worm's protein during the course of an infection would be more reliable than wet weight as an indicator of growth.

Attempts to cultivate Acanthocephala *in vitro* have met with little success, but must be continued if the influence of the environment on growth is to be understood.

CARBOHYDRATE METABOLISM

Carbohydrate metabolism has been studied more intensively than any other aspect of helminth physiology (von Brand, 1966). The original stimulus to these studies may have been the simple yet accurate methods developed for determining glycogen content. Intestinal helminths, including acanthocephalans, have been found to contain relatively large amounts of glycogen (von Brand, 1966); up to 13 % of the dry weight of *M. hirudinaceus* may be glycogen. The histochemical localization of this polysaccharide has been studied in *M. hirudinaceus* (von Brand, 1939*a*), in *Leptorhynchoides thecatus* and *Echinorhynchus coregoni* (von Brand, 1939*b*, 1940), in *E. coregoni*, *E. gadi*, *Pomphorhynchus bulbocolli*, *Neoechinorhynchus cylindratus* and *N. emydis* (Bullock, 1949*b*) and in *P. minutus* (Crompton, 1965). In all these species glycogen is found associated with the musculature and body wall, which may contain as much as 80 % of the glycogen of the worm (von Brand, 1939*a*).

Among acanthocephalans, most work on carbohydrate metabolism has been done on *M. dubius* maintained *in vitro*. Recently, *M. dubius* was found to catabolize glucose, under either aerobic or near anaerobic conditions, to ethanol and carbon dioxide (Ward and Crompton, 1969). The evidence indicated that the process was very similar to alcoholic fermentation by yeast. From a review of some of the literature on carbohydrate catabolism, these authors concluded that the carbohydrate metabolism of large helminths is characterized by the formation, via pyruvate, of acidic and other end-products independently of oxygen, a process indicating that the glycolytic pathway is involved. Large helminths in this context are those which are relatively large compared with their environment.

This generalization, which is considered to apply to parasites *in vitro*, was devised from the following lines of evidence.

(1) Adult acanthocephalans appear to have high rates of glucose utilization *in vitro*.

(2) The presence of glucose extends the period of spontaneous movement *in vitro* when worms are maintained in saline under anaerobic conditions.

(3) Glucose is catabolized to organic acids and other end-products in both the presence and absence of oxygen (table 7).

(4) Glycolytic enzymes have been extracted from adult *M. hirudinaceus* (Dunagan and Scheifinger, 1966*b*), but key enzymes of the citric acid cycle could not be extracted from this parasite (Dunagan and Scheifinger, 1966*a*).

(5) Intermediate metabolites of glycolysis have been extracted from adult *M. dubius* (Graff, 1964, 1965; Bryant and Nicholas, 1965) and *P. minutus* (Crompton and Lockwood, 1968), but some metabolites associated more especially with the citric acid cycle have not been found.

The generalization does not appear to be supported by all the results on carbohydrate metabolism of adult acanthocephalans. These anomalous results, however, are difficult to interpret for parasites living *in vivo*. The uptake of oxygen by *M. dubius in vitro*, in the presence of a variety of carbohydrates, was measured by Laurie (1959). No obvious function can be given to the absorbed oxygen which does not appear to have been involved in carbohydrate catabolism since acidic excretory products were formed (table 7) and Laurie gives no evidence of carbon dioxide evolution. Bryant and Nicholas (1966) also studied oxygen uptake by whole *M. dubius*, by pieces and by a particulate preparation. For worms kept without glucose in equilibrium with air, males used about 0·28 ml/h per gram of tissue and females used 0·19 ml/h per gram of tissue. Succinate was found to stimulate the oxygen uptake of the pieces and par-

TABLE 7. *Substances reported to be excretory products of acanthocephalans studied* in vitro

	Excretory substance									
	Carbon dioxide	Ethanol	Acetate	*n*-Butyrate	Formate	Propionate	*n*-Valerate	Lactate	Succinate	Unidentified
A *Moniliformis dubius* (Laurie, 1957, 1959)	.	.	√	.	√	.	.	√	.	.
M. dubius (Ward and Crompton, 1969)	√	√	√	√	√	√	√	√	√	.
P *Polymorphus minutus* (Crompton and Ward, 1967)	.	.	.	.	.	.	.	√	√	√
E *Neoechinorhynchus* spp. (Dunagan, 1964)	.	.	.	.	.	.	.	√	.	.

ticles, but not the whole worms. Succinate is an excretory product of *M. dubius* (Ward and Crompton, 1969) and is not likely to be reabsorbed and metabolized by healthy whole worms. These technical details and results may prove to be important clues in explaining the observation that many intestinal helminths, which are usually assumed to be adapted to life independent of oxygen *in vivo*, absorb oxygen when it is available *in vitro*. This topic is discussed below (p. 45).

SOME RELATIONSHIPS BETWEEN THE ENVIRONMENT AND CARBOHYDRATE METABOLISM

The availability of carbohydrate and the effect of starvation

When rats, infected with *M. dubius*, are maintained on a carbohydrate-free diet, the growth of the worms is reported to be arrested (Read and Rothman, 1958). This finding, which depends on the reliability of using wet weight as a measure of growth, shows that an acanthocephalan cannot obtain sufficient energy for protein synthesis unless carbohydrate is present in its environment. The worms' potential ability to grow is not lost during carbohydrate deprivation and, on addition of starch to the host's diet, their growth rate recovers. Complete starvation of a host for 48 h, however, causes a fall of the glycogen level of *M. dubius* to 12% of its normal value (Read and Rothman, 1958) and expulsion of *P. minutus* from ducks (Nicholas and Hynes, 1958). The fact that infections are not destroyed by carbohydrate deprivation lasting several days, indicates that worms can metabolize food reserves and other nutrients to provide energy for feeding, osmo- and ionic regulation and the maintenance of their position. It is difficult to assess the respective importance of food reserves and environmental nutrients because any food eaten by the host promotes intestinal secretions (p. 21) which may be used by a parasite to obtain energy and conserve stored food. A starving host provides an environment more detrimental to an acanthocephalan than a host eating an abnormal diet because starvation reduces all the nutrients available in intestinal secretions.

Carbohydrate, mediated through the host's diet, may be assumed to be essential as the energy source for growth in all acanthocephalans, but lack of food for the host appears to affect worms from homeothermic hosts only. For example, *Neo-*

echinorhynchus emydis from turtles and *A. ranae* from frogs can live for several months in fasting hosts (p. 25). Although it may be argued that starvation is unnatural and injurious while fasting is normal and controlled, other evidence suggests that fundamental differences exist between the metabolism of worms from homeotherms and those from poikilotherms. Thus, worms from poikilotherms can be kept alive *in vitro* much longer than those from homeotherms. *Pomphorhynchus bulbocolli* from fresh-water fish developed for 72 days *in vitro* (Jensen, 1952) and *Neo-echinorhynchus* spp. from turtles lived for 96 days (Dunagan, 1962), but *P. minutus* (Crompton and Ward, 1967) and *M. dubius* (Nicholas and Grigg, 1965) became moribund after 19 and 8 days respectively.

Glucose

There need be little doubt that glucose is a nutrient of prime importance for all adult Acanthocephala. Glucose is metabolized by species from all three orders of the phylum (table 6), polymerized glucose, in the form of starch given to starving rats infected with *M. dubius*, causes a recovery of that worm's glycogen level (Read and Rothman, 1958) and glycogen itself, which is another form of polymerized glucose, is the major carbohydrate reserve (p. 40). Furthermore, free glucose has been detected in the environment of *P. minutus* in ducks (Crompton, 1966).

The availability and concentration of glucose in the environment affects glucose uptake and catabolism. The estimate of Crompton and Lockwood (1968) that glucose is required by *P. minutus in vivo* at the rate of 4 μg/h per milligram of worm means that a mature female worm weighing 5 mg (fig. 11) would use 20 μg glucose/hr. If all this glucose were catabolized by glycolysis, $1{\cdot}33 \times 10^{17}$ molecules of ATP would be produced hourly for synthesis, muscle contraction and so on.

Not all the glucose absorbed is used immediately for energy and some is stored as glycogen. This process may occur irrespective of the environmental glucose concentration if the worm's glycogen is in a state of dynamic equilibrium. Nothing is known about the concentration of glucose needed to maintain any acanthocephalan's glycogen level, but some evidence suggests that a diurnal rhythm of glycogenesis exists in *M. dubius* infecting laboratory rats (Read and Rothman, 1958). More glucose may be present in the worm's environment by night if the hosts are less active than in the day. Wild rats, however, are more active at night and the rhythm may be reversed.

Oxygen and glucose

More energy can be obtained from glucose catabolized with the participation of oxygen than from glucose catabolized independently of oxygen. The relationship between carbohydrate catabolism and oxygen in intestinal helminths has been discussed by von Brand (1966) and Ward and Crompton (1969), and the latter authors concluded that acanthocephalans *in vivo* probably obtain energy from glucose independently of oxygen. This conclusion was made in spite of the fact that, at a recent symposium, Fenn (1964) and Chance, Schoener and Schindler (1964) suggested that an oxygen tension of 1 mm Hg within the mitochondrion of a recognized aerobic organism was sufficient to permit ATP production linked to the electron flow between reduced co-enzymes and oxygen. One reason why adult acanthocephalans may not rely on oxygen, in spite of its presence in their environments (table 4), is the probable inability of the gas to diffuse quickly enough through the tissues to reach the mitochondria at a tension of 1 mm Hg. This possibility appears to be supported by the fact that Bryant and Nicholas (1966) observed a greater oxygen uptake with pieces and particles of *M. dubius* than with whole worms (p. 43). The statement by Smith (1969) that he could find no evidence of cytochromes in *Neoechinorhynchus* spp. from the intestine of mullet, *Mugil cephalus*, favours the view that acanthocephalans obtain energy independently of oxygen. Nevertheless, the participation or otherwise of oxygen in energy production by acanthocephalans *in vivo* will continue to be a theoretical problem until experiments can be undertaken with the worms in their environments. The fact remains that oxygen uptake by acanthocephalans *in vitro* has been measured (Laurie, 1959; Bryant and Nicholas, 1966), but no satisfactory explanation for the process has been offered. The excretion of acidic and other end-products by the parasites in the presence of oxygen rebuts the view that the gas is involved in energy production by normal aerobic processes.

Some recent work on brewers' yeast, *Saccharomyces cerevisiae*, described by Linnane (1965), suggests that environmental and experimental glucose concentrations may control the uptake and use of oxygen. Cells of *S. cerevisiae* grown aerobically contain mitochondria with cristae and the full complement of enzymes associated with electron transport, oxidative phosphorylation and the citric acid cycle. Cells of *S. cerevisiae* grown anaerobically are devoid of mitochondria, but mitochondrial synthesis is induced by the presence of oxygen. Mitochondria produced in this way, in the yeast cells grown originally under anaerobic conditions, resemble the mitochondria of yeast cells grown aerobically in both morphological and enzymic characteristics. The synthesis of certain mitochondrial enzymes, however, has been found to be repressed by glucose. Concentrations of 0·6 % glucose do not prevent enzyme synthesis, but those higher than 5 % glucose effect inhibition.

This relationship between oxygen, glucose and mitochondrial activity can be used to explain several puzzling results obtained from investigations of the carbohydrate metabolism of acanthocephalans. For example, the mitochondria from the muscles of *P. minutus*, depicted in the paper by Crompton (1965), may have relatively prominent cristae because oxygen is diffusing into the muscles of the worm situated close to the intestinal villi, but the worm may still catabolize carbohydrate independently of oxygen because the high environmental glucose concentration inhibits enzyme synthesis (table 4). Conversely, the oxygen uptake reported for worms *in vitro* could have resulted from the presence of oxygen with either no glucose or too low a concentration to repress enzyme synthesis. The excretion of acidic end-products may have been due to the worms' using endogenous carbohydrate, which may be equally repressive above certain concentrations, or exogenous carbohydrate in the incubation vessels while enzymes were being synthesized. Further examples, particularly from other types of helminth, could be cited in support of this hypothesis, which awaits experimental test.

Carbon dioxide and glucose

Carbon dioxide is available in the environment of many helminths. The nematode, *Heterakis gallinae*, fixes CO_2 and so reduces its rate of utilization of carbohydrate (Glocklin and Fairbairn, 1952). Most helminths have now been found to fix CO_2; in the cestode, *Hymenolepis diminuta*, Scheibel and Saz (1966) have obtained evidence to show that CO_2 is joined to pyruvate and the resulting C_4 compound is reduced to succinate and excreted. It is probable that reduction of the C_4 compound involves the regeneration of NAD or NADP and results in the formation of one molecule of ATP for every molecule of CO_2 fixed. The formation of ATP in this way could explain the carbohydrate-sparing effect of CO_2 fixation reported by Glocklin and Fairbairn.

The fixation of CO_2 by adult *M. dubius* has been studied by Graff (1965), who incubated worms in a saline containing $NaH^{14}CO_3$, but no glucose. His results are in agreement with those of Scheibel and Saz, in that the route for fixation must have involved either pyruvate or phosphoenol pyruvate. The radioactive label was detected in aspartic acid, alanine, serine, malate, succinate, fumarate, lactate, pyruvate and oxaloacetate. No evidence of CO_2 fixation by *M. dubius* was found by Ward and Crompton (1969), even when *M. dubius* was incubated in a sealed vessel containing gaseous CO_2 of its own production. This apparent contradiction may derive from the fact that Ward and Crompton included glucose in their incubation medium. Thus, yet another feature of the carbohydrate metabolism of adult acanthocephalans may be assigned to glucose. It now appears that acanthocephalans fix CO_2 when exogenous glucose is unavailable, but not otherwise. This interpretation is supported by the work of Kilejan (1963), who found that *M. dubius* could synthesize the same amount of glycogen from glucose *in vitro* in either the presence or absence of CO_2. Glycogenesis from glucose by the cestode, *H. diminuta*, on the other hand, was negligible *in vitro* unless CO_2 was present (Fairbairn *et al.* 1961), and Kilejan confirmed this result. These findings on CO_2 fixation by *M. dubius* and *H. diminuta* are of special interest because they demonstrate a physiological difference between two helminths which appear to require the same environment (p. 22).

Trehalose

The presence of trehalose may prove to be an important ecological factor for the acanthocephalan worms which develop in insect intermediate hosts and complete their life cycles in final hosts which regularly eat insects. Trehalose is now known to be the common blood sugar of insects (Gilmour, 1965) and it is obvious that hosts which eat insects will introduce trehalose into the environment of adult acanthocephalans. Evidence for the metabolism of trehalose by *M. dubius* was obtained by Fairbairn (1958), who detected this disaccharide in extracts, and by Fisher (1964), who studied the synthesis of trehalose by tissue minces. Further critical work should be done to determine whether trehalose is absorbed by *M. dubius* or not. A positive result would indicate that it could be a nutrient of special significance to *M. dubius* in wild rats whose intestines probably contain trehalose from ingested insects. On the other hand, the trehalose detected in the tissues of adult acanthocephalans may result from the conversion of glucose to trehalose as a device to facilitate glucose absorption when this sugar is available in the environment. The conversion of glucose to trehalose in the worm would tend to maintain a concentration gradient in favour of the entry of glucose into the worm. The observation of Dunagan (1962), that trehalose prolonged the survival of *Neoechinorhynchus* spp. *in vitro*, suggests that this worm can absorb it.

Amino acids

The work of Graff (1964, 1965) has shown that ^{14}C from radioactive glucose and from radioactive $NaHCO_3$ is incorporated into alanine, aspartic acid and serine in *M. dubius*, and Crompton and Lockwood (1968) also noted that radioactive amino acids were present in extracts of *P. minutus* which had been incubated *in vitro* with ^{14}C glucose. In the worms' environments, however, free amino acids are usually available and the worms are more likely to use glucose for energy production rather than amino acid synthesis. This point could be resolved by incubating worms with ^{14}C glucose and amino acids in the proportions found in the environment.

Bile salts

Since bile salts have been shown to inhibit the carbohydrate metabolism of the cestode *H. diminuta in vitro* (Rothman, 1958), it is likely that *M. dubius* and other acanthocephalans will be similarly affected. Rothman studied the effects of the sodium salts of cholic, taurocholic and glycocholic acids at concentrations similar to those in the rat's intestine. The detergent bile salts may have disrupted the absorptive mechanisms in the surface membrane of the cestode; an acanthocephalan would also be susceptible to this type of damage. In contrast, inhibition of a worm's carbohydrate metabolism by bile salts presumably does not normally occur *in vivo*. The release of bile into the intestine is synchronized with the arrival of food; worms such as *M. dubius*, living near the bile duct, may be protected by the combination of bile salts with undigested, dietary lipid to form emulsified droplets and micelles. *Polymorphus minutus* and acanthocephalans from posterior environments, cannot be protected in this way and in fact tolerate bile acids at a concentration of one-twentieth of that in bile from the host's gall bladder (Crompton and Nesheim, 1970).

SOME EFFECTS OF ACANTHOCEPHALAN CARBOHYDRATE METABOLISM ON THE ENVIRONMENT

The excretory products of acanthocephalans will affect their environments (table 7). Succinate and lactate can be expected to be absorbed and metabolized by the host. Succinate may be assumed to enter directly into the host's citric acid cycle and lactate will be oxidized to pyruvate and then metabolized with a gain of 18 molecules of ATP. If lactate is utilized in this way, the host will obtain 36 molecules of ATP, rather than 38 molecules, for each molecule of glucose absorbed by the parasite and converted by it to lactate. This fact reinforces the conclusion of von Brand (1966) that intestinal helminths rarely deprive their hosts of significant amounts of food. Some acanthocephalans contain so much glycogen that they may be acting as storage tissue for their hosts.

The effects of the ethanol excreted by *M. dubius* have been discussed by Ward and Crompton (1969). Some ethanol may be metabolized by the host's intestinal flora, though the environment of *M. dubius* will not contain many micro-organisms (table 4) and most ethanol, therefore, will be absorbed by the rats. In mammals, the nervous system, vision, neuro-muscular co-ordination and the cutaneous blood vessels are known to be affected by ethanol and, consequently, rats infected with *M. dubius* may be at a serious disadvantage compared with uninfected members of the population to which they belong. It is known that, when rats have received daily oral doses of ethanol, slices of their kidneys metabolize ethanol more efficiently than slices from rats that have not been dosed (Leloir and Muñoz,

1938). The results show that dosing the rats produces a tenfold increase in alcohol dehydrogenase activity (Dixon and Webb, 1964). Recently, rats receiving ethanol daily for a considerable period of time, have been discovered to possess increased alcohol dehydrogenase activity in their intestinal walls (Mistilis and Birchall, 1969). The significance of this result, for rats infected with *M. dubius*, is that much of the ethanol excreted by the parasite may never be absorbed into the host's general circulatory system.

It has been found by Holmes (1961) that, in cases of concurrent infections by *M. dubius* and the cestode, *H. diminuta*, the latter is displaced considerably to a more posterior position (fig. 4*e*). As long as *M. dubius* and *H. diminuta* are in the same environment, their physical contacts may result in the ethanol excreted by the acanthocephalan affecting the co-ordination of the cestode. Consequently, the cestodes would relax and be moved by peristalsis while *M. dubius* would remain in the preferred environment.

SUMMARY

Acanthocephalans infecting poikilothermic hosts appear to have fundamentally different mechanisms for withstanding environmental conditions produced by starvation of their hosts than do worms infecting homeotherms. The evidence suggests that, *in vivo*, acanthocephalans catabolize carbohydrate independently of oxygen and that glucose is degraded to pyruvate by the glycolytic pathway. Glucose is an important source of energy for acanthocephalans and its environmental concentration may control glycogenesis, oxygen utilization, mitochondrial development and the fixation of carbon dioxide. The end-products of carbohydrate metabolism which are excreted by acanthocephalans may be metabolized by the host or be effective in displacing competitors.

Insufficient information about protein and lipid metabolism is available to promote a discussion of these subjects.

CHAPTER 5

Reproduction

REPRODUCTION in the Acanthocephala is characterized by a complicated, and at present little understood, copulatory procedure followed by a mechanism for the discharge of mature eggs by the female worms. This selective release of eggs ensures that any egg eaten by the correct intermediate host is likely to develop into a cystacanth infective to another final host. However, before copulation and egg production can be considered, the longevity, morphology and sexual dimorphism of the worms must be discussed.

SOME FEATURES OF MALE AND FEMALE WORMS

Sex ratio and longevity

Studies on adult *Polymorphus minutus* in ducks have demonstrated that a sex ratio of 1:1 exists for the first part of the infection (Nicholas and Hynes, 1958; Crompton and Whitfield, 1968*b*), and this situation probably applies to all species of Acanthocephala. This ratio of 1:1 for young adult worms must have arisen during the parasites' development in their intermediate hosts. The sex ratio of *Echinorhynchus truttae* developing in *Gammarus pulex* is 1:1 irrespective of the sex and age of the host and of the size of the infection (Parenti, Antoniotti and Beccio, 1965). The sex ratio of Acanthocephala, therefore, appears to be under direct genetic control and is not influenced by the environment provided by the intermediate host.

If the ratio had been larger in favour of females, the inference would have been that males mated with more than one female, a feature which could lead to the production of a greater number of infective eggs. On the other hand, the chances of males becoming established in final hosts would be smaller than those for females since fewer male cystacanths would be available for infection. Thus, through a ratio of 1:1, natural selection has favoured the regular production of moderate numbers of eggs rather than the more erratic production of vast numbers.

Difficulty may be encountered in determining the sex ratio from natural infections of acanthocephalans because females are usually observed to be more numerous than males. For example, Chubb (1964), who studied *Echinorhynchus clavula* (= *Acanthocephalus clavula*) in fresh-water fish, reported finding that females were more abundant. These observations result from the fact that male worms do not usually live as long as females and, consequently, the sex ratio changes during the course of an infection. When ducks are infected at a known time with cystacanths of *P. minutus*, the worms are found in their environment in approximately equal numbers for the first 16 days of the infection (Crompton and Whitfield, 1968*b*). Thereafter, the rate of loss of males is about 4 times that of females and no males remain after 36 days although some females continue to live in the host for 20 more days. Similarly, Awachie (1966) reported that male *E. truttae* begin to be discharged from the trout's intestine more rapidly than females 45 days after the start of the infection.

The conclusion that female acanthocephalans live longer than males appears to apply in cases where males have copulated; but males which have not copulated may live longer than those which have. This qualification arises from studies on infections of *Macracanthorhynchus hirudinaceus* in pigs (Kates, 1944). Four pigs were infected and, apart from recording the longevity and numbers of the sexes present, the duration of egg production by the females was determined. In one pig, which had been infected for over a year, two females, whose period of egg production was over, and one male were recovered. The presence of the male worm is difficult to explain unless it had not mated, particularly since Kates observed that other male worms were expelled from the host earlier in the course of the infection. The significance in nature of the increased longevity of unmated males would be that, in superimposed infections, newly established females would have increased chances of being fertilized. Data are not available to permit discussion of the longevity of unfertilized female worms.

Sexual dimorphism and morphology

Apart from the numbers and arrangement of proboscis hooks, the most obvious external difference between the sexes is that of size, female worms usually being larger than males. This statement probably applies to fertilized female worms and their increase in size reflects the growth of their bodies to accommodate the eggs which in turn add to the weight of the females. The differences between the weights of male and female *P. minutus* are illustrated in fig. 11 (p. 39) and it can be seen that the weight of females begins to increase rapidly a few days before males begin to leave the host from the sixteenth day of the infection onwards. The expulsion of males provides more space in the environment both for the growing females and for the establishment of new worms in superimposed infections. The despatch of the male worms may be an environmental effect mediated through the host or else through the female worms. It would be interesting to extend the observations of Graff and Allen (1963), who found that male *Moniliformis dubius* grew larger in rats in the absence of females, and discover (i) how female *M. dubius* repress the growth of the males, (ii) whether this repression is experienced by both unmated males and males which have mated and (iii) to establish whether similar mechanisms exist for other species of Acanthocephala.

The morphology of the male and female reproductive systems is highly characteristic of Acanthocephala and general descriptions for many species are to be found in Meyer (1933), Petroschenko (1956, 1958) and Yamaguti (1963) and in the first chapter of this book (p. 5). Recently, thorough studies of the female and male reproductive systems of *P. minutus* have been made by Whitfield (1968 and 1969 respectively). These descriptions and his review of earlier work demonstrate the general similarity between all known acanthocephalan reproductive systems.

The female reproductive system or efferent duct system provides the route for the entry of spermatozoa and for the discharge of eggs. On copulation, spermatozoa pass from the vagina, up the muscular lower uterus and anterior uterus to the uterine bell from which they escape into the pseudocoelom amongst the ovarian balls, which originally formed the ovary within the ligament. The ovarian balls then release fertilized eggs and, as these develop, they are sorted by the uterine bell, which directs mature eggs down the uterus to the vagina while the immature eggs are returned to the pseudocoelom to develop further. The sorting function of the uterine bell has long been assumed because mature eggs only are found in the uterus and host's faeces, while eggs at all stages of development are found in the female's pseudocoelom. The morphology of the efferent

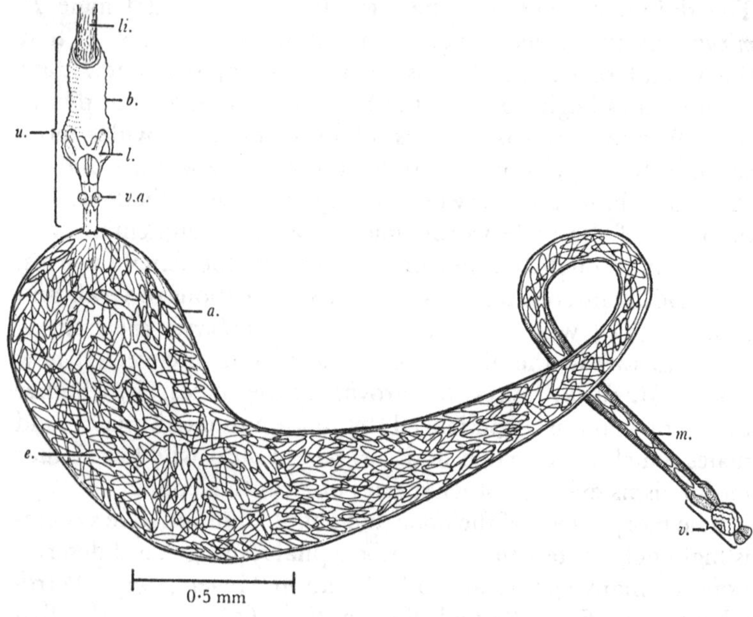

Fig. 12. The mature efferent duct of female *Polymorphus minutus* (after Whitfield, 1968, fig. 1). *a.*, Anterior uterus; *b.*, bell wall syncytium; *e.*, eggs; *l.*, lateral pocket syncytium; *li.*, ligament; *m.*, muscular lower uterus; *u.*, uterine bell; *v.*, vagina; *v.a.*, ventral accessory cell.

duct system of *P. minutus* and a representation of how the uterine bell achieves egg sorting are shown in figs. 12 and 13.

The male reproductive system is characterized by the presence of an extensible bursa which is everted, by contractions of its own musculature, to grasp the female during copulation (fig. 2*b*, p. 4). Saefftigen's pouch is also considered to be involved in the movements of the bursa. Some acanthocephalans—for example, *Telosentis molini* (Van Cleave, 1923)—have small, genital spines round their posterior ends. These spines occur in

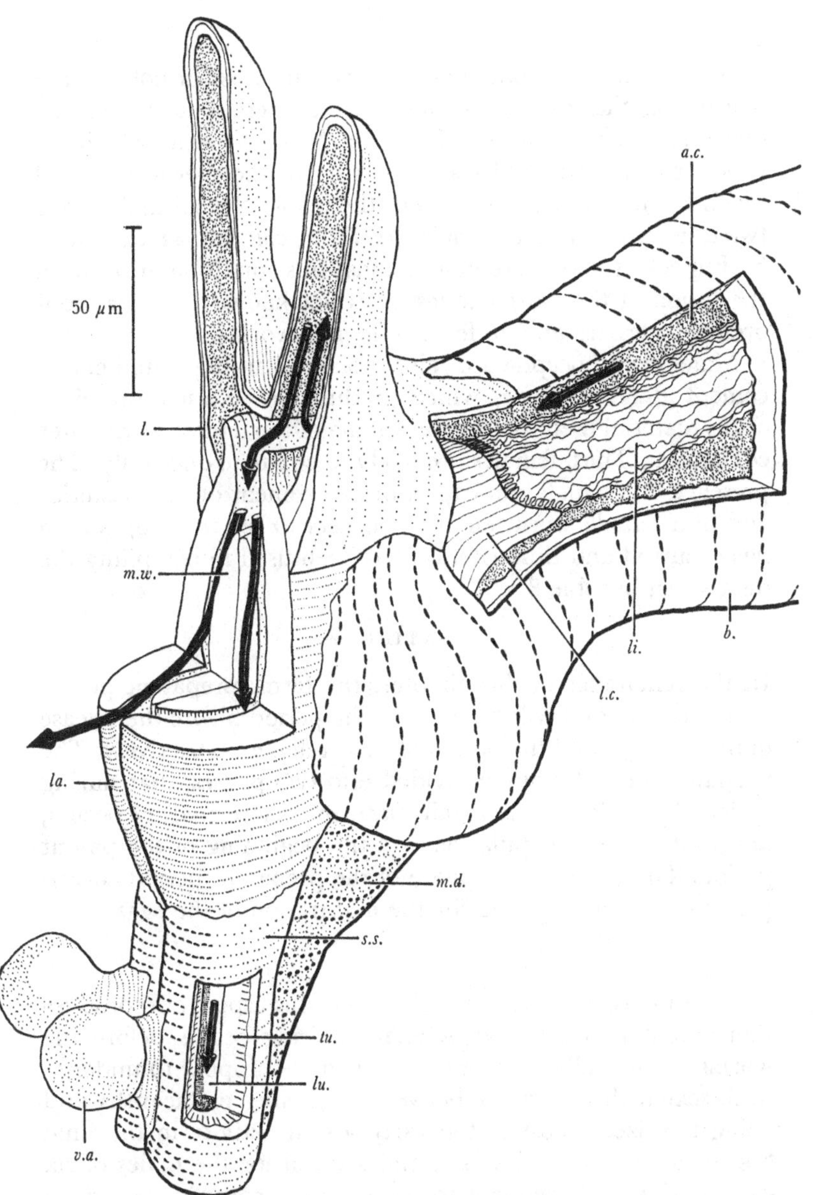

Fig. 13. Stereogram of mature uterine bell of *Polymorphus minutus*, cut away to reveal complex internal luminal system. Possible routes for egg translocation indicated by heavy arrows (after Whitfield, 1968, fig. 9). *a.c.*, Anterior chamber of uterine bell; *b.*, bell wall syncytium; *l.*, lateral pocket syncytium; *la.*, lappet; *l.c.*, ligament attachment cushion; *li.*, ligament; *lu.*, lumen of uterine duct tube; *m.d.*, median dorsal wall; *m.w.*, median wall cell; *s.s.*, sheathing syncytium; *tu.*, tube of uterine duct cell; *v.a.*, ventral accessory cell.

both sexes and are sometimes believed to assist in holding the worms together during copulation, but there is no direct evidence for their having this function. Mature sperm, which are produced in the paired testes, pass down the sperm duct and reach the female system through the penis. Not all males have two testes and 14 of 208 male *Acanthocephalus jacksoni*, examined by Bullock (1962), were monorchids; this condition arises from the fusion of the paired testes. It is not known whether monorchid males are able to fertilize female worms.

Male acanthocephalans also possess varying numbers of cement glands (fig. 2*b*) which produce an adhesive proteinaceous material used to plug the vagina of the female worm after copulation. Thus, females are likely to copulate once only. The presence of a copulation cap of brown cement on the posterior end of a female worm enables workers to see that copulation has occurred and this criterion has been used in compiling the data given in table 8.

MATING

In the remainder of this chapter, the terms prepatent period and patent period will be used for the period before the release of mature eggs and the period of their release, respectively. The prepatent period may be divided into two phases, that during which the male worms reach maturity before mating occurs, and that between copulation and the beginning of the patent period. Thus, the second phase of the prepatent period corresponds to the time needed for the maturation of the eggs.

Pairing

Little information is available about this aspect of reproduction and it is not known whether pairing begins before the worms are sexually mature or not. Females appear to undergo little sexual development between becoming cystacanths and being fertilized; most of the early sexual activity in the final host concerns the males. The times taken for the males of six species of Acanthocephala to reach sexual maturity are given in table 8.

Acanthocephalan worms face several hazards when pairing takes place. Unless a host contains a very heavy infection, both worms or one of them will have to move to find the mate or to come close enough for copulation. Circumstantial evidence in

TABLE 8. *The course of reproduction in Acanthocephala*

Species	Final host	Prepatent period phase 1 (up to mating)	Prepatent period phase 2 (up to egg-maturation)	Total duration of prepatent period	Patent period (egg release)	References
A *Macracanthorhynchus hirudinaceus*	Mammal	—	—	8 weeks	10–12 months	Kates (1944)
A *Moniliformis dubius*	Mammal	3 weeks[1]	3 weeks[2]	6 weeks	18 weeks[3]	[1] Robinson (1965) [2] Nicholas (1967) [3] Sita (1949)
P *Echinorhynchus truttae*	Fish	3 days	8–9 weeks	$8\frac{1}{2}$–$9\frac{1}{2}$ weeks	3 weeks	Awachie (1966)
P *Leptorhynchoides thecatus*	Fish	4 weeks	4 weeks	8 weeks	—	DeGiusti (1949*a*)
P *Pomphorhynchus bulbocolli*	Fish	3 weeks	7 weeks	10 weeks	—	Jensen (1952)
P *Polymorphus minutus*	Bird	5 days	3 weeks	$3\frac{1}{2}$ weeks	3–4 weeks	Crompton and Whitfield (1968*b*)
E *Octospinifer macilentis*	Fish	?8–10 weeks	?6–8 weeks	?16 weeks	—	Harms (1965)

the literature suggests that males move more than females, as, for example, in the case of *P. minutus* (Nicholas and Hynes, 1958). Males are more frequently found free in the intestine than females, and the smaller tissue response made by the intestinal wall against the proboscis of males supports the view that males are more active. Movement for an acanthocephalan must occur in spite of peristalsis, rhythmic segmentation and pendular movements (p. 16), all of which cause the intestinal contents to rotate in a spiral, anti-clockwise manner (Crompton, 1969). It is not known how partners find each other, whether one sex produces attracting chemicals which excite the other or whether pairing is the result of random movements. Nor is it known how far acanthocephalans move; individual *P. minutus* may be separated by 30 cm and individual *M. hirudinaceus* by 300 cm.

Copulation

This process has rarely been observed and, since numerous records exist of adult acanthocephalans being found in the intestines of vertebrates, it may be concluded that copulation is either a rapid process or else one which occurs at night. In fact, in a host resting at night, less material may be present in the intestine and movement will be easier for the worms. The argument that copulation would be disrupted on the death of the host is not substantiated at present because the best description of the process is for *Acanthosentis dattai* taken from the intestine of freshly killed fish in India by Podder (1938).

Copulation in *A. dattai* was observed on several occasions between worms obtained in July and August. The everted bursa of the male was seen to become attached to the posterior end of the female and, by a single sucking mechanism, the female's vagina was drawn out and held in position in the cavity of the bursa. The bursa was then described as being partially retracted and it was assumed that at this point the penis was inserted and the spermatozoa were transferred. These observations are supported by the sketches depicting males and females of *Sphaerechinorhynchus rotundocapitatus* (Johnston and Deland, 1929) and *Gorgorhynchus clavatum* (Yamaguti, 1963) *in copula*. Copulation in some species of Acanthocephala could now be investigated in detail since it has been reported to occur *in vitro* with *Neoechinorhynchus emydis* and *N. pseudemydis* (Dunagan, 1962).

The second phase of the prepatent period consists of the maturation of the eggs; details for several species of Acanthocephala are given in table 8. Maturation involves the development of the acanthor larva and formation of the egg-shells, both processes occurring in the pseudocoelom of the female under conditions defined by the physiological responses of the female to the environment. Since little is known about the environment provided by the pseudocoelom for egg development, this subject will not be discussed, but the literature on morphological aspects of embryology has been reviewed by Nicholas and Hynes (1963).

The patent periods of four species of Acanthocephala are listed in table 8; this information has been obtained from examinations of hosts' faeces for mature eggs. The time when eggs are first released by females depends on the time of copulation, which is, in its turn, affected by the size of the population of worms.

A detailed study of the egg production of *M. hirudinaceus* in pigs showed that the average maximum daily egg production of a female worm was about 260,000 eggs (Kates, 1944). An earlier and less rigorous study suggested that the same worm produced 82,000 eggs per day (Wolffhügel, 1924). Variations are bound to occur when estimating egg production and Kates pointed out that his results applied to a worm during the peak of its patent period and that Wolffhügel's figure was probably an average value for the whole of the patent period. Even if a female *M. hirudinaceus* releases only 82,000 eggs per day, it will produce about 23 million mature eggs during the patent period of 10 months.

The release of eggs by *P. minutus* reaches a maximum rate of 2000 eggs per day (Crompton and Whitfield, 1968*b*). Eggs from the body cavity of an 18-day-old female worm are infective to the intermediate host, although eggs have never been found in ducks' faeces until the worms are 22 days old. The significance of this observation is not yet understood, but it is possible that the uterus must be full of mature eggs before it contracts and discharges them into the host's intestine.

SUMMARY

Male and female Acanthocephala of the same species develop in the intermediate host to give a sex ratio of 1 : 1. This ratio appears to be maintained until after copulation has occurred in the final host. Male worms are subsequently expelled earlier than females. Apart from living longer than males which have mated, females grow larger, in order to accommodate the developing eggs. The uterine bell, which sorts mature eggs from immature ones, and the muscular uterus control the release of mature eggs from the female worm. The means by which mature worms find each other before mating, the frequency and manner of copulation and the cause of the expulsion of male worms before females, are at present matters of speculation rather than fact.

CHAPTER 6

The Egg and the Infection of the Intermediate Host

ARTHROPODS become infected with acanthocephalans by ingesting acanthor larvae enclosed in shells. These larvae arise from cleavage of fertilized eggs in the pseudocoelom of the female worm. Once the development of the acanthor is well advanced, the shell is produced round it within the original vitelline membrane of the egg. Many authors call this infective stage the shelled acanthor or shelled embryo, while others refer to the egg. Since, by definition, an egg is a mature female germ-cell, the terms 'shelled acanthor' and 'shelled embryo' are more accurate, but the term 'egg' has been retained in this chapter as the most convenient term for descriptive purposes.

Until the egg is eaten by the correct intermediate host, it may be considered as a resting stage capable of enduring at least three different environments. These include the intestine and rectum of the final host, the conditions in water or on land where eggs have been deposited with faeces of the final host and the alimentary tract of the intermediate host. The fact that acanthocephalan eggs can survive a variety of conditions and yet respond rapidly to those which stimulate hatching is one of the most interesting aspects of acanthocephalan physiology.

THE MORPHOLOGY AND STRUCTURE OF THE EGGS

The structure of the shell must contribute to the survival of infective eggs (fig. 14). The shells of archi- and palaeacanthocephalans contain four layers, while those of eoacanthocephalans contain three or four. These layers have been labelled numerically in fig. 14, rather than naming them fertilization membrane and so on (West, 1964). It seems to be unwise to use terms of special significance to mammalian embryologists for layers of unknown function in the acanthocephalan egg. Also, any nomenclature for these layers should be withheld until they have been studied with the electron microscope.

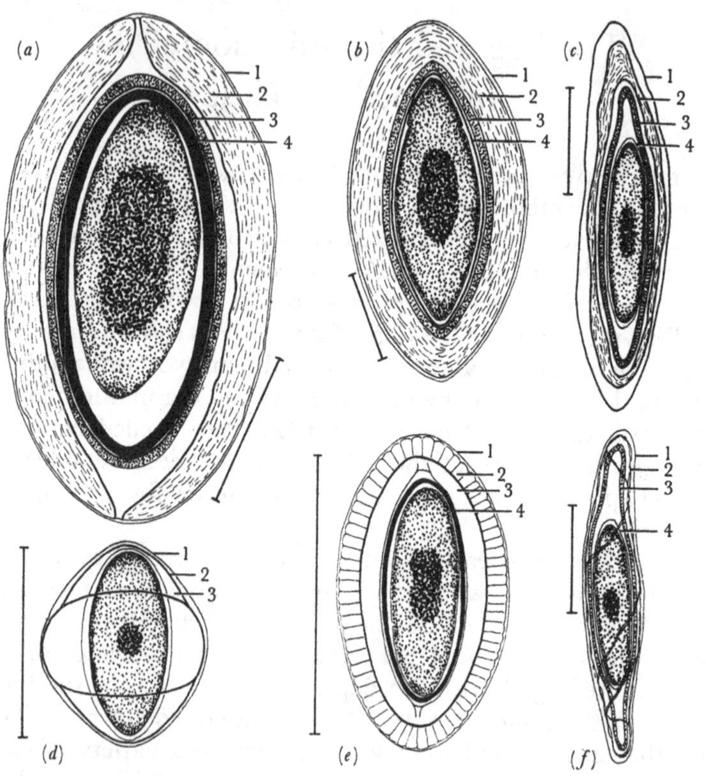

Fig. 14. Eggs of six species of Acanthocephala. (*a*) *Macracanthorhynchus hirudinaceus* (after Meyer, 1933, fig. 369); (*b*) *Moniliformis dubius* (after Nicholas, 1967, fig. 3E); (*c*) *Echinorhynchus gadi* (after West, 1964, fig. 1); (*d*) *Neoechinorhynchus emydis* (after Hopp, 1954, fig. 3); (*e*) *Neoechinorhynchus rutili* (after Meyer, 1931, fig. 27); (*f*) *Acanthocephalus jacksoni* (after West, 1964, fig. 2). The egg shells are numbered 1, 2, 3 and 4 for convenience only and not to indicate homologies between eggs. Scale represents 25 μm.

The shells have been reported to be composed of protein, chitin, keratin-like material, polyphenols, cellulose, elastin-like material and fibrin (West, 1963; Monné, 1964). The location of all these materials in the layers is not known, but good evidence for the presence of chitin in either layers 3 or 4 or both of them has been obtained from *Macracanthorhynchus hirudinaceus* (von Brand, 1940), *Profilicollis botulus* and *Polymorphus minutus* (Monné and Hönig, 1954), *Echinorhynchus gadi* and *Acanthocephalus jacksoni* (West, 1963) and *Moniliformis dubius* (Edmonds, 1966). The shells of these species contain material resistant to prolonged treatment with hot, concentrated alkali. The evidence for the presence of keratin-like material applies to *P. botulus* and *P. minutus*, where layer 2 is dissolved by solutions of sodium sulphide and sodium thioglycollate (Monné and Hönig, 1954). These compounds attack any material containing disulphide linkages and, until electron micrographs and X-ray diffraction pictures have been obtained, layer 2 is better considered as containing protein stabilized by disulphide linkages.

All the acanthors described to date are basically similar and are equipped with hooks or blades at the anterior end. These structures, which are called the rostellar apparatus, are operated by tiny muscles and are believed to assist the acanthors in boring into the intestinal tissues of their intermediate hosts. The rostellar apparatus could possibly be involved in cutting through the shell when the acanthor escapes on hatching, and so be analogous to the egg-bursters of certain insects. Unless the acanthors are in a state of anabiosis before being stimulated to hatch, the permeability of the shell, however slight, will expose them to some of the conditions prevailing in the environments which have to be endured. Consequently, an acanthor itself must also be tolerant of varying conditions.

THE EGG AS A RESTING AND RESISTANT STAGE

Some information on the duration of the infectivity of eggs is given in table 9. Most of the species listed parasitize aquatic hosts in the temperate regions of the world, so that the egg may be the vital stage by which such parasites survive the winter. The eggs of acanthocephalans from terrestrial hosts may be exposed to conditions which fluctuate even more than those in water. For example, some eggs of *M. hirudinaceus*, contained in pig faeces on experimental plots of soil in Maryland, U.S.A., were found to retain their infectivity to the intermediate host, *Cotinus nitida*, for $3\frac{1}{2}$ years (Spindler and Kates, 1940).

TABLE 9. *Periods for which acanthocephalan eggs have been stored in water without loss of infectivity*

Species	Temperature (°C)	Period	Reference
Archiacanthocephala			
*Moniliformis dubius**	5	4 weeks	Edmonds (1966)
Palaeacanthocephala			
Leptorhynchoides thecatus	4	9 months	DeGiusti (1949*a*)
Polymorphus minutus (syn. *P. magnus Bezvbik*, 1957)	10–17	6 months	Petroschenko (1956)
Pomphorhynchus bulbocolli	4	6 months	Jensen (1952)
*Prosthorhynchus formosus**	5	10 months	Schmidt and Olsen (1964)
Eoacanthocephala			
Octospinifer macilentis	4	9 months	Harms (1965)
Neoechinorhynchus rutili	4	6 months	Merritt and Pratt (1964)
Paulisentis fractus	4	Many months	Cable and Dill (1967)

* Infects terrestrial intermediate hosts.

These eggs had withstood all the climatic and physical conditions occurring on the surface of soil. After this preliminary finding, the environmental factors thought likely to damage the eggs were simulated in the laboratory by Kates (1942), who exposed eggs to them. His experiments and work are summarized in table 10. Under natural conditions, exposure to sunlight is probably the most serious hazard, though many eggs may fall into sheltered cracks and depressions at the soil's surface.

The literature also contains several observations illustrating the resistant properties of infective eggs. A small proportion of eggs of *P. minutus* was found to be infective to *Gammarus pulex* after being frozen for 2½ months at −22 °C (Hynes and Nicholas, 1963). When eggs of *M. hirudinaceus* were mixed with pigeon food, eaten by pigeons and recovered from the birds' faeces, they were still able to infect larval beetles of the genus *Phyllophaga* (Glasgow and DePorte, 1939). Similarly, eggs of *Moniliformis clarki* retained their infectivity after passing through the entire length of the alimentary tract of the final host, *Peromyscus maniculatus sonoriensis* (Crook and Grundmann, 1964).

TABLE 10. *Summary of the work of Kates* (1942) *on the viability of eggs of* Macracanthorhynchus hirudinaceus

	Conditions to which eggs were exposed	Results
1	To temps of 36° to 90 °C for 10 min.	Viability lost at 60 °C and above
2	(*a*) Freezing in water for 140 days in temps. from −10° to −16 °C	Viability retained like that of controls
	(*b*) Freezing when dry for 140 days in temps from −10° to −16 °C	Viability retained like that of controls
3	(*a*) Drying in air for 50 days at temps of 5 to 9 °C, 37° to 39 °C	Viability retained like that of controls
	(*b*) Drying in air for 265 days at temps of 21° to 26 °C	Viability retained like that of controls
4	(*a*) Alternate wetting and drying on soil at 37° to 39 °C	Viability destroyed after 368 days' treatment
	(*b*) Alternate wetting and drying on soil at 2° to 5 °C	Viability retained like that of controls after 551 days' treatment
5	Irradiation with UV light for 10 min. when eggs spread in thin layer	Viability lost

These findings show that the eggs can withstand hydrochloric acid, pH changes, bile acids and the variety of enzymes of the vertebrate digestive system.

INFECTION OF THE INTERMEDIATE HOST

Infection of the intermediate host consists of the hatching of the egg and the penetration of the host's intestinal tissues by the acanthor. The environment where this takes place is very often the alimentary tract of an insect or crustacean. Details of the morphology and physico-chemical conditions in the alimentary tracts of *Periplaneta americana* and *Gammarus* spp. are described below and summarized in figs. 15 and 16; these arthropods are the intermediate hosts of *M. dubius* and *P. minutus*, respectively. The discussion, which is confined to the fore- and midguts, probably applies to all the orthopteroid insects and amphipod crustaceans listed in table 1.

Some features are common to both environments. For example, the foregut of all arthropods is ectodermal in origin and lined with cuticle, while the midgut is endodermal and devoid of cuticle. The peristaltic contractions of the muscular

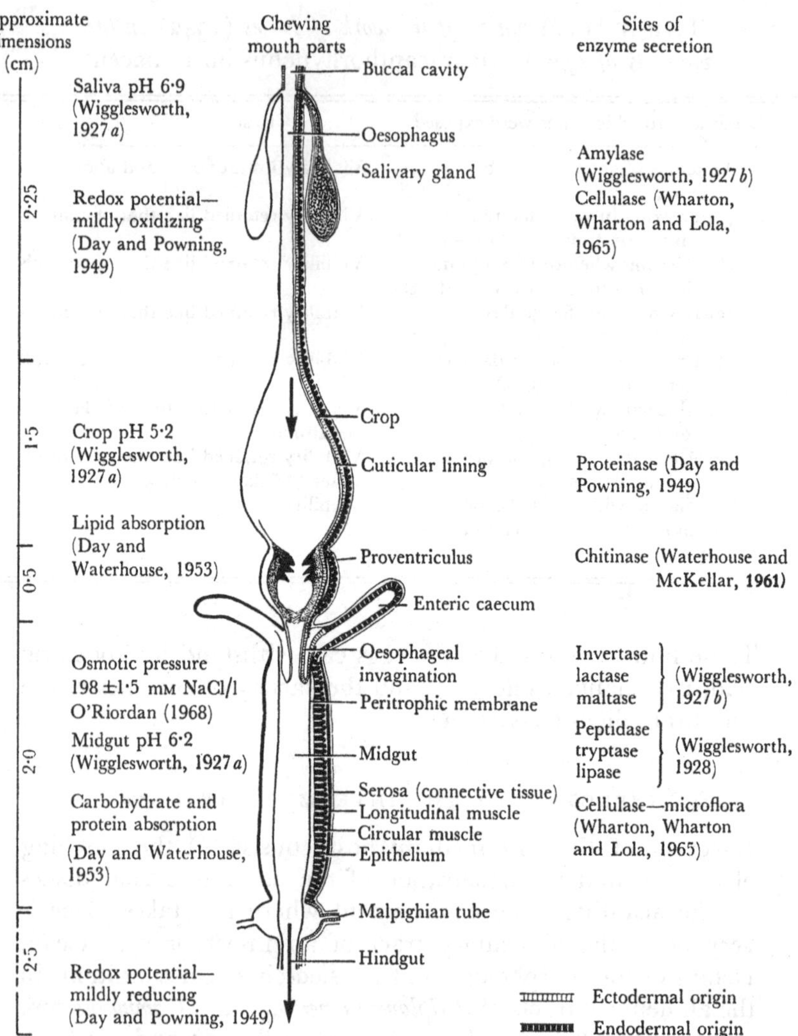

Fig. 15. Diagrammatic representation of the alimentary tract of *Periplaneta americana* as an environment of eggs and acanthors of *Moniliformis dubius*.

coat of the intestinal wall propel food through the intestinal tract.

There are, however, several differences between these environments in *P. americana* and *Gammarus*. Cockroaches are erratic feeders and the crop is a distensible organ with functions

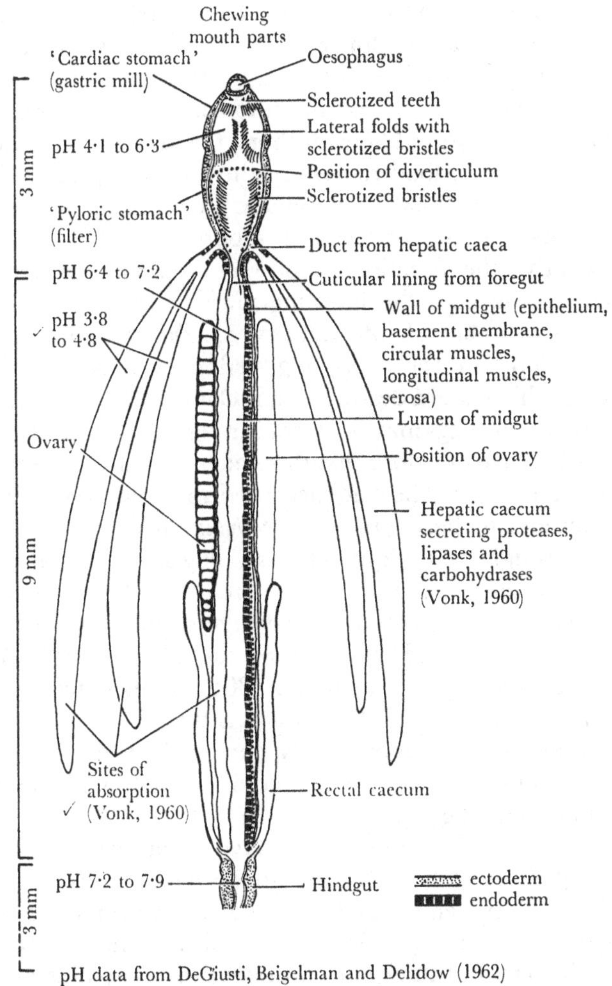

Fig. 16. Diagrammatic representation of the alimentary tract of *Gammarus* as an environment of eggs and acanthors of *Polymorphus minutus*.

of digestion and food storage (Day and Waterhouse, 1953). Digestion is effected by copious amylytic secretions from the salivary glands and regurgitated enzymes from the midgut (Cornwell, 1968), but food storage results from the availability of food and the feeding behaviour of the insect. In cockroaches maintained in laboratories, a meal usually takes about 24 h to pass through the tract, but most food will have left the midgut

$2\frac{1}{2}$ h after ingestion (Snipes and Tauber, 1937). When a cockroach has been starved for 2 days, food enters the midgut within 5 min of ingestion, the midgut lumen is filled 15 min later and remains in this condition for another 100 min (Day and Powning, 1949). These authors noted that, in spite of starvation, traces of food remain in the crop for 3 days. Other authors claim that food may remain in the crop for several days (Abbott, 1926) or even 2 months (Sandford, 1918). Consequently, the functioning of the crop and the rate of food passage through the intestine must be important factors in the establishment of infections of *M. dubius*. If the food is stored for long in the crop, ingested eggs may experience conditions which could destroy them. Conversely, if the food is stored briefly, the eggs may be swept through the intestine before they have hatched or before acanthors have had time to penetrate the intestinal tissues. The evidence cited in table 11 shows that most acanthors, including those of *M. dubius*, penetrate the host's intestine through the midgut. The hatching stimulus must activate the acanthor in the foregut before the midgut is reached.

Gammarus, in contrast, are scavengers, feeding whenever opportunity arises and no part of the alimentary tract is modified as a storage organ (Sars, 1867). Thus hatching may be a better ordered process for *P. minutus* than for *M. dubius*, but the acanthors may be more easily damaged by the bristles of the gastric mill (fig. 16) than those of *M. dubius* by the proventricular teeth (fig. 15).

Once the proventricular teeth have been negotiated, the acanthor of *M. dubius* must go through the peritrophic membrane of chitinous fibrils embedded in a matrix of protein (Mercer and Day, 1952; Peters, 1969). This hazard to infection occurs in many insects, but may not be present in many crustaceans. Some copepods (Gauld, 1957) and decapods (Forster, 1953) possess a peritrophic membrane like that of insects, but in *Gammarus*, the contents of the intestinal lumen are separated from the epithelium by a gelatinous substance which may not be as tough a barrier as a laminated membrane. Another hazard to infection, involving the peritrophic membrane, is the likelihood of damage to free acanthors from abrasive particles of roughage which the membrane restricts to the intestinal lumen.

TABLE II. *Some observations on the hatching of eggs in the alimentary tracts of normal intermediate hosts*

Species	Site and time of hatching after ingestion	Observer's remarks	Reference
Archiacanthocephala			
Macracanthorhynchus hirudinaceus	Midgut	—	Kates (1943)
M. ingens	Midgut	Acanthor probably plays an active part in escaping from the egg	Moore (1946*b*)
Mediorhynchus grandis	Midgut 24 to 48 h	—	Moore (1962)
Moniliformis clarki	Foregut	Hatch in three stages: outer layer swells and breaks anteriorly, middle layer breaks anteriorly, acanthor escapes	Crook and Grundmann (1964)
M. dubius	Midgut 24 to 48 h	—	Moore (1946*a*)
	Midgut 2 to 6 h	—	Edmonds (1966)
Palaeacanthocephala			
Echinorhynchus truttae	Intestine (? midgut) 1 to 20 h	Hatching achieved by action of host's gastric mill, host's enzymes and movements of acanthor	Awachie (1966)

TABLE II. (*cont.*)

Species	Site and time of hatching after ingestion	Observer's remarks	Reference
Leptorhynchoides thecatus	$\frac{3}{4}$ h	Hatching requires host's digestive enzymes and movements of the acanthor	DeGiusti (1949*a*)
Polymorphus minutus	Midgut 1 to 20 h	Polar ends of the eggs bitten off within 5 min of ingestion	Hynes and Nicholas (1957)
Prosthorhynchus formosus	Midgut $\frac{1}{4}$ to 2 h	Two outer layers of the eggs spring open along preformed ridges; acanthor penetrates the inner layer and escapes	Schmidt and Olsen (1964)
Eoacanthocephala			
Neoechinorhynchus emydis	Intestine ?(midgut), 8 to 12 h	—	Hopp (1954)
N. rutili	Intestine (? midgut), 6 to 12 h	—	Merritt and Pratt (1964)
Octospinifer macilentis	Intestine (? midgut) within 4 h	Outer layer splits transversely at the equator of the egg and the acanthor enclosed by the other layers becomes motile	Harms (1965)
Paulisentis fractus	? within 4 h	—	Cable and Dill (1967)

Observations on hatching in vivo

Some observations on hatching recorded in the literature are presented in table 11. No information is available about the numbers of eggs ingested by a host, the number which hatch or the number which eventually become cystacanths, but hundreds of cystacanths have been found in some hosts (Miller, 1943). In such cases, many acanthors must have penetrated the midgut and the fact that the hosts did not die, either directly from mechanical damage or indirectly from invasion of the haemocoele by intestinal micro-organisms, may be attributed to the healing properties of the midgut epithelium. This tissue, in adult *P. americana*, is replaced every 40 to 120 h (Day and Powning, 1949), and it heals completely when wounded, whereas the fore- and hindguts do not (Day, 1952). Thus, the evolution of infection of arthropods by acanthocephalans has resulted in the parasite damaging the tissue which the host can best repair. Occasionally, however, hosts have been reported to be killed by invading acanthors from laboratory infections in which no control was exercised over the number of eggs eaten (DeGiusti, 1949*a*; Crompton, 1964*b*; Crook and Grundmann, 1964).

Inspection of table 11 shows that the time for hatching to occur, after the eggs have been ingested, is very variable. This variability may merely reflect the periods which workers allowed to pass between feeding the hosts and looking for the acanthors, or it may be genuine and result from the retention of food and eggs in storage organs. The point is illustrated by the different times of hatching of eggs of *M. dubius* in *P. americana*, given as 24 to 48 h by Moore (1946*a*) and 2 to 6 h by Edmonds (1966). If 2 h is assumed to be the shortest time for the eggs of *M. dubius* to hatch (Edmonds) it can be inferred that they can persist for 46 h (Moore). The hatching time of 2 h fits closely with the observation that food leaves the intestine between 2 and 2½ h after ingestion when cockroaches are fed under laboratory conditions (p. 68).

The observations in table 11 also indicate that, after the hatching stimulus has been received, the acanthors are active in escaping from their shells, at least from the innermost layer (fig. 14, layer 4). The data in this table confirm the conclusion of Meyer (1933) that preformed raphes occur in some of the layers of the shell and facilitate the release of the acanthors.

Observations on hatching in vitro

An investigation of the hatching of acanthocephalan eggs *in vitro* was undertaken by Edmonds (1966) on *M. dubius*, and it will be profitable to consider his findings in conjunction with the conditions in the alimentary tract of *P. americana* (fig. 15) and the information in table 11.

Since nematode eggs are stimulated to hatch, in certain cases, by the combination of concentrations of undissociated and dissolved carbon dioxide, and the hydrogen ion concentration and redox potential of their environments, Edmonds assumed that similar factors could activate acanthocephalan eggs. He went on to assess the hatching of eggs incubated in saline in controlled conditions; some of his results are shown in fig. 17. The eggs were found to hatch when the molarity was greater than 0·2 (fig. 17*a*) and hatching was increased by bicarbonate ions rather than phosphate ions (fig. 17*b*). Hatching was minimal below pH 7·0 except when promoted by the presence of 1 % carbon dioxide in the gas phase (fig. 17*c*). The presence or absence of reducing substances, which would have affected the redox potential, had no significant effect.

The factors which appear to stimulate hatching *in vitro* actually occur in the alimentary tract of the cockroach (fig. 15). The molarity of the crop contents is known to be 0·2 (Treherne, 1957), hydrogen ion concentrations of pH 6·0 and over exist (Wigglesworth, 1927*a*) and carbon dioxide and bicarbonate ions are sure to be present.

Having obtained evidence that the eggs of *M. dubius* respond to stimuli received from the environment, Edmonds examined the possible role of host chitinase in hatching. Chitinase occurs in the alimentary tract of cockroaches (fig. 15) and its substrate occurs in the acanthocephalan egg (p. 63). This enzyme from cockroaches, and other sources, had no detectable effect on hatching, although chitinase was found to be produced by acanthors in response to the hatching stimuli. At present, the function of the chitinase has not been determined, but penetration of the inner layers of the shell and the host's peritrophic membrane are obvious possibilities. The acanthors were also considered by Edmonds to cut their way out of the shell. These observations and some of those in table 11 demonstrate that acanthors actively escape from the shell after receiving hatching stimuli.

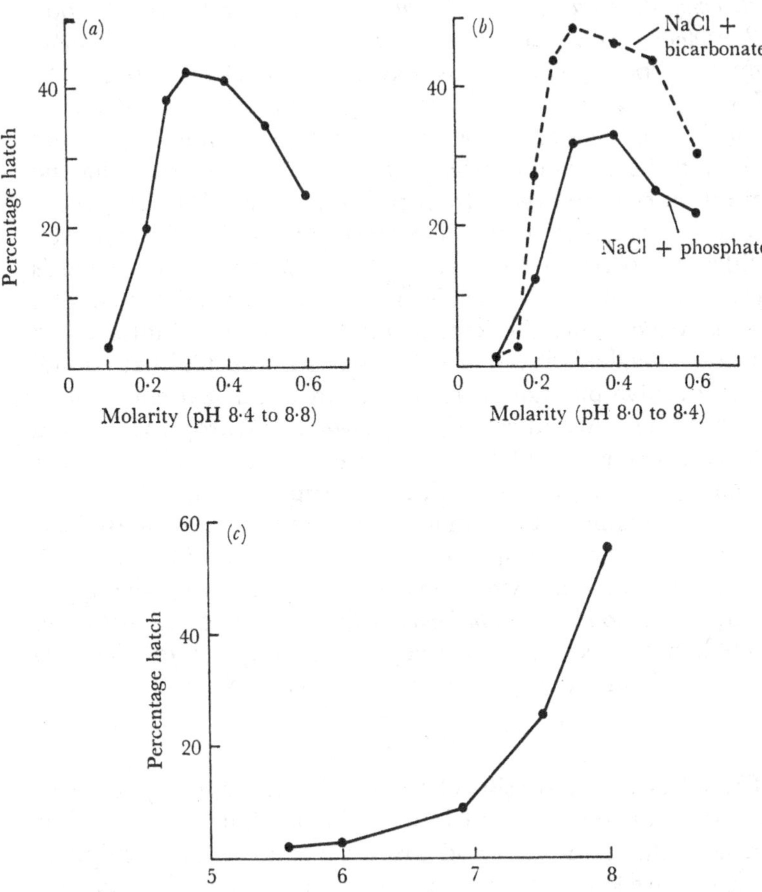

Fig. 17. The hatching of eggs of *Moniliformis dubius in vitro*. (*a*) The effect of molarity on percentage hatch; (*b*) the effect of bicarbonate ions on percentage hatch; (*c*) the effect of pH on percentage hatch. Ordinate represents the percentage hatch. (After Edmonds, 1966, figs 1 to 4).

Other work on hatching *in vitro* has been carried out by Manter (1928), who described a method for hatching eggs, and by Moore (1942, 1946*b*), who modified Manter's technique. The procedure consisted of removing eggs from the pseudocoelom of a mature female worm, drying them and then rewetting, whereupon some hatched. Hatching was induced for eggs of *M. hirudinaceus*, *M. ingens*, *Mediorhynchus grandis*,

Hamanniella tortuosa and *M. dubius*. Eggs of *Centrorhynchus* spp., *P. minutus*, *Neoechinorhynchus emydis*, *N. cylindratus* and *Prosthorhynchus formosus* tested in this way did not hatch (Manter, 1928; Moore, 1942; Hynes and Nicholas, 1957; Schmidt and Olsen, 1964). This procedure of drying and re-wetting eggs is not likely to be of significance in the life cycle because it has no effect on eggs recovered from the hosts' faeces (Manter, 1928; Moore, 1942). Thus, in some species, there is a fundamental difference between infective eggs recovered from the female's pseudocoelom and those which have passed through the efferent duct system. The mechanism of hatching by Manter's and Moore's methods is not understood. It is possible that stimuli from the manipulations of the procedure could have contributed to hatching. Acanthors of *N. cylindratus* (Ward, 1940*a*) and *N. rutili* (Merritt and Pratt, 1964) can be released from their shells by gentle pressure with a coverslip.

The remaining observations on hatching *in vitro* are difficult to interpret. For example, eggs of *Prosthorhynchus formosus* hatch within 15 to 20 min when mixed with crushed digestive gland from the isopod, *Armadillidium vulgare* (Schmidt and Olsen, 1964), but those of *P. minutus* do not respond to pieces of *Gammarus*' intestine (Hynes and Nicholas, 1957).

SUMMARY

The infective egg consists of an acanthor larva, equipped with hooks or blades, enclosed in a shell, containing three or four layers. Chitin, protein and protein stabilized by di-sulphide linkages are known to be constituents of the shell. The eggs are very resistant to many conditions, but respond quickly to hatching stimuli in the foregut of the intermediate host. The evidence indicates that the molarity, hydrogen ion concentration and carbon dioxide content of the foregut constitute the hatching stimuli for eggs of *M. dubius*. These stimuli activate the acanthors, which produce chitinase, cut their way out of the shell and leave the intestinal lumen through the midgut epithelium. Examination of the literature suggests that this description of infection applies to most species of Acanthocephala.

CHAPTER 7

The Arthropod Haemocoele and the Development of Acanthocephala

ALL the acanthocephalans which have been studied in detail develop in an arthropod intermediate host. For most species, the arthropod serves as the only intermediate host involved in the life cycle, but reliable evidence has shown that *Neoechinorhynchus emydis* grows in the gasteropod *Campeloma rufum*, as well as in the ostracod *Cypria maculata*, before becoming infective to the final host (Hopp, 1954).

Development begins as soon as the acanthor moves out of the intestinal lumen of the intermediate host. Several species—for example, *Polymorphus minutus* and *Macracanthorhynchus hirudinaceus*—may be associated with their host's intestinal tissues for a considerable time, while others—for example, most eoacanthocephalans—are associated with the host's intestine for a few hours only. All acanthocephalans, however, undergo their major developmental changes in their host's haemocoele. Those acanthocephalans which are associated with the intestinal wall are invariably observed to be enveloped in a transparent capsule by the time they have become free in the haemocoele.

Since the development of *P. minutus* has been studied in detail by Greef (1864), Hynes and Nicholas (1957), Crompton (1964*b*) and Butterworth (1969*a*) and that of *Moniliformis dubius* by Yamaguti and Miyata (1942), Moore (1946*a*), Sita (1949) and King and Robinson (1967), an analysis is presented below of the environments of these species in the haemocoeles of their intermediate hosts. *Polymorphus minutus* develops in six species of the amphipod crustacean, *Gammarus* including *G. pulex*, while *M. dubius* develops in the orthopteroid insect, *Periplaneta americana*. Nine out of the twenty-two species of Acanthocephala, listed in table 1 (p. 8), are also found in amphipods and orthopteroid insects.

THE ENVIRONMENT OF *POLYMORPHUS MINUTUS* AND *MONILIFORMIS DUBIUS* IN THE HAEMOCOELES OF THEIR INTERMEDIATE HOSTS

The basic component of the environment is the haemolymph, consisting of haemocytes and plasma, which contains electrolytes and many organic molecules. The temperature of the haemolymph is not controlled, but other factors—for example, osmotic pressure, amino acid concentrations and water content—are regulated (Buck, 1953; Lockwood, 1968) and thus acanthocephalans develop in precise conditions which are essential to their host's physiological processes. During the course of evolution of the host–parasite relationship, the acanthocephalan may have become so dependent on these conditions that they now govern the specificity of a parasite for its host. There may also be a tendency for completely parasitic organisms like acanthocephalans, which must develop in an arthropod and reproduce in a vertebrate, to be under evolutionary pressure to become better adapted to the environment in the intermediate host. One result of such specialization could be the attainment of sexual maturity within the arthropod host and the omission of the vertebrate host. If this position were ever reached during the aeons of the future, parasitism by acanthocephalans would have returned to what is believed to be its archetypal condition.

Physical factors

Conditions in the crustacean and insect haemocoeles reveal fundamental differences between the environments. For instance, crustacean haemolymph has the respiratory function of oxygen transport while the tracheal system effects this in insects, although some oxygen is bound to be dissolved and carried in an insect's haemolymph (Buck, 1953). The concentration of amino acids is higher and more variable in the haemolymph of insects than in crustaceans. This variation of amino acid concentration in insects is involved in the regulation of osmotic pressure (Sutcliffe, 1963), and could result in a less stable food supply for acanthocephalans in insects compared with those in crustaceans. The composition of the haemolymph differs from species to species, and several authors (Florkin, 1960; Wyatt, 1961; Florkin and Jeuniaux, 1964) have listed inorganic ions, organic acids, enzymes, hormones and other proteins in a wide variety of insects and crustaceans. This work serves to show that differences exist not only between the haemocoeles of crustaceans and insects, but also between those of species of either class. Furthermore, conditions in the haemocoele vary for an individual host at different stages in its moulting cycle (see below).

Some of the features of the environments of *P. minutus* and *M. dubius* are cited in table 12. These data, however, apply to hosts in the intermoult period; conditions will fluctuate considerably during moulting. *Gammarus*, for example, which lives for about fifteen months, undergoes fourteen moults in the first half of its life and six moults in the second (Kinne, 1959). These are average figures for moulting frequency which may be affected by temperature, nutrition, wounding and salinity. Moulting in decapod crustaceans involves an uptake of water which increases the animal's haemolymph volume to 2½ times the intermoult value; this uptake of water is facilitated by a rise in the haemolymph's osmotic pressure due to the absorption of calcium and magnesium salts from the endocuticle (Lockwood, 1968). Similar events probably occur in *Gammarus* and, since *P. minutus* requires 60 days for development at temperatures as high as 17 °C (Hynes and Nicholas, 1957), the parasite must adjust to the changes associated with moulting on at least two occasions. Acanthocephalans such as *M. dubius*, developing in adult insects, avoid moulting changes, but those in nymphs and larvae do not.

Biotic factors

The principal biotic factor in the environment is the population of the host's haemocytes, which are responsible for the defence of the body against metazoan parasites (Salt, 1963). The haemocytes of insects have been reported to form about 1 % of the haemolymph volume (Buck, 1953) and the physiological state of the insect has been considered to affect the number of haemocytes present in the haemocoele (Wigglesworth, 1959). In *P. americana*, adult insects contain about 110,000 ± 17,000 haemocytes/mm^3 (Wheeler, 1963); most of these will be phagocytic cells (Wigglesworth, 1959). The haemocyte count rises before moulting, a situation suggesting that haemocytes are involved in engulfing cytolytic material

TABLE 12. *Some features of the haemocoeles of* Gammarus *and* Periplaneta, *the environments of* P. minutus *and* M. dubius *respectively*

	G. pulex	*P. americana*
Haemolymph volume	1·27±0·25 μl/mg dry weight ∴ 12 μl in a 50 mg living amphipod (Butterworth, 1968*a*)	*ca.* 20% of wet weight (Wheeler, 1963) 140 μl in a 800 mg living insect (Cornwell, 1968)
Haemolymph osmotic pressure	120 to 142 mM-NaCl/l (Butterworth, 1969*b*) 150 mM-NaCl/l (Lockwood, 1961)	202 mM-NaCl/l (O'Riordan, 1968)
Haemolymph composition		
Water	*ca.* 95%	90% (Buck, 1953)
Electrolytes	See Florkin (1960), Lockwood (1968)	See Buck (1953), Florkin and Jeuniaux (1964)
Carbohydrates	20 to 37 mg glucose/100 ml (Butterworth, 1968*b*)	*ca.* 880 to 1400 mg trehalose/100 ml (Wyatt, 1967)
Amino acids	0·5 mg/100 ml (Butterworth, 1968*b*)	40 mg to 385 mg/100 ml (Pratt, 1950)
Lipids	181 mg/100 ml (Butterworth, 1969*b*)	See Chino and Gilbert (1965)
Carotenoids	3 mg/100 ml (Butterworth, 1969*b*)	
Haemolymph pH	7·5 (inference from Waterman, 1960)	7·4 (O'Riordan, 1968)
Dissolved gases		
Oxygen	*ca.* 2·5 ml/100 ml (inference from Lockwood, 1968)	See Buck (1953)
Carbon dioxide	*ca.* 30 ml/100 ml (inference from Wolvekamp and Waterman, 1960)	See Buck (1953)

produced during absorption of the endocuticle (Tauber, 1937; Wheeler, 1963). Thus, increase in phagocytic haemocytes during moulting is another factor which could make moulting a potentially dangerous period for the developing acanthocephalan (p. 86). Similar information about the haemocytes of crustacean hosts is not available, but there is no reason to suppose that these haemocytes are different from those of insects; they have already been shown to be responsible for defence reactions (Crompton, 1967).

Little is known about the effects of micro-organisms, protozoan infections and other metazoan parasites on acanthocephalans developing in the haemocoele of arthropods. In *Gammarus*, *P. minutus* can develop successfully in multiple infections, although the rate of development is retarded (Crompton, 1964*b*), in superimposed infections (Hynes and Nicholas, 1958) and in the presence of the acanthocephalan, *Echinorhynchus truttae* (Awachie, 1967). Other organisms—for example, the yeast, *Cryptococcus gammari* (Goodrich, 1928), and the trematode, *Crepidostomum farionis* (Baylis, 1931)—inhabit the haemocoele of *Gammarus*, and these are potential competitors of *P. minutus*. Few studies have been undertaken on the interactions between helminths in an arthropod host. *Gammarus pulex*, *P. minutus* and some of the parasites mentioned above would provide a good experimental system for study, particularly since the host's cuticle is transparent.

DEVELOPMENT OF ACANTHOCEPHALA

Four distinct stages, the early, middle and late acanthellae and the cystacanth, are recognizable during an acanthocephalan's development. This terminology is taken from a recent study of the development of *P. minutus* in *G. pulex* by Butterworth (1969*a*). The early acanthella is characterized by the appearance of the proboscis and genital primordia from the central nuclear mass, the establishment of the axis of development and the growth and deployment of the cortical nuclei (fig. 18*a*, *e*). *Polymorphus minutus*, which develops for some time in the intestinal tissues, becomes free in the haemocoele during this stage. The middle acanthella begins with the appearance of the pseudocoelom, includes proboscis and muscle development and usually ends with the fragmentation of the giant cortical nuclei (fig. 18*b*, *c*). Butterworth considered that the middle acanthella could be subdivided into phases I and II. Her observations, however, were made at a temperature of 10 °C, with the result that the parasites developed very slowly, encouraging the recognition of stages which would have passed unobserved at higher temperatures. The late acanthella is marked by the inversion of the proboscis and neck, the completion of the lemnisci and the specialization of the body wall (fig. 18*d*). When growth and development have stopped and the parasite is infective to the final host, it is called the cystacanth, which is described in chapter 9.

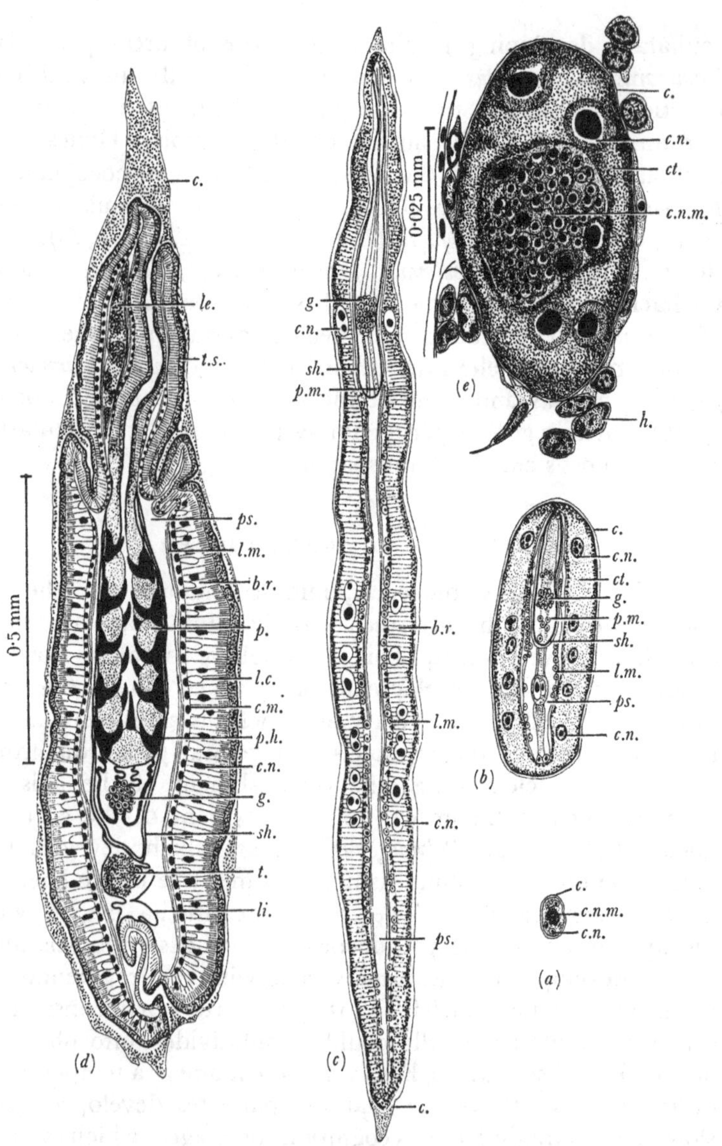

Fig. 18. The morphology of the developmental stages of *Polymorphus minutus*. (*a*) Early acanthella stage; (*b*) middle acanthella stage I; (*c*) middle acanthella stage II; (*d*) late acanthella stage; (*a*), (*b*), (*c*) and (*d*) are drawn to the same scale. (*e*) early acanthella stage drawn to a higher magnification than that shown in (*a*). *b.r.*, radial layer; *c.*, capsule; *ct.*, cortex; *c.m.*, circular muscle; *c.n.*, cortical nucleus; *c.n.m.*, central nuclear mass; *g.*, ganglion; *h.*, haemocyte; *le.*, lemniscus; *li.*, ligament; *l.c.*, lacunar channel; *l.m.*, longitudinal muscle; *p.*, proboscis; *ps.*, pseudocoelom; *p.h.*, proboscis hook; *p.m.*, proboscis retractor muscle; *sh.*, proboscis sheath *t.*, testis; *t.s.*, trunk spine.

Since arthropods are poikilotherms, the rate of development of acanthocephalans varies according to the ambient temperature. Thus, *P. minutus* reaches the cystacanth stage after 60 days at 17 °C (Hynes and Nicholas, 1957) and after 150 days at 10 °C (Butterworth, 1969*a*). The proportion of time spent in each of the stages, however, remains constant and approximately 40%, 26% and 34% of the developmental time of *P. minutus* is taken up with the early, middle and late acanthellae respectively. Since it is sometimes difficult to decide when an acanthor becomes an early acanthella, the time spent as an acanthor has been included with that spent as an early acanthella in these estimations. This constant relative duration of the developmental stages applies to all the acanthocephalans for which morphological descriptions exist, provided that the stages are defined according to Butterworth. The development of thirteen species of Acanthocephala is illustrated in this way in fig. 19. Any description, therefore, of acanthocephalan development can be assumed to be respresentative for the group and, in addition to the references cited in fig. 19, the description of Nicholas (1967) gives a clear summary of development.

Development of the body wall

The body wall of the developing acanthocephalan is an interesting tissue because it becomes specialized to form the absorptive surface and major metabolic tissue of the adult worm while retaining these functions for the parasite in the intermediate host. In many cases, parts of the body wall become adapted to protect the cystacanth stage from mechanical damage during the infection of the final host (p. 94). It is, therefore, surprising that the literature did not contain a detailed study of the development of the acanthocephalan body wall until that of *P. minutus* was described by Butterworth (1969*a*).

In the early acanthella, the cortex is simple, the globular-shaped giant nuclei have prominent nucleoli and there are few visible organelles. The surface of the early acanthella is bounded by a plasma membrane penetrated by small pores beneath which is a layer thought to be composed of small vesicles. During this phase of development, the number and size of lipid bodies in the body wall appears to increase. In the middle acanthella, the cortical nuclei become arranged in annuli, while the numbers of mitochondria and golgi bodies increase and more endoplasmic reticulum develops. These organelles become concentrated below the layer of vesicles in a manner indicative of intense synthetic activity. Eventually, the arrangement of the cortical nuclei breaks down and the lacunar channels appear in the part of the cortex destined to

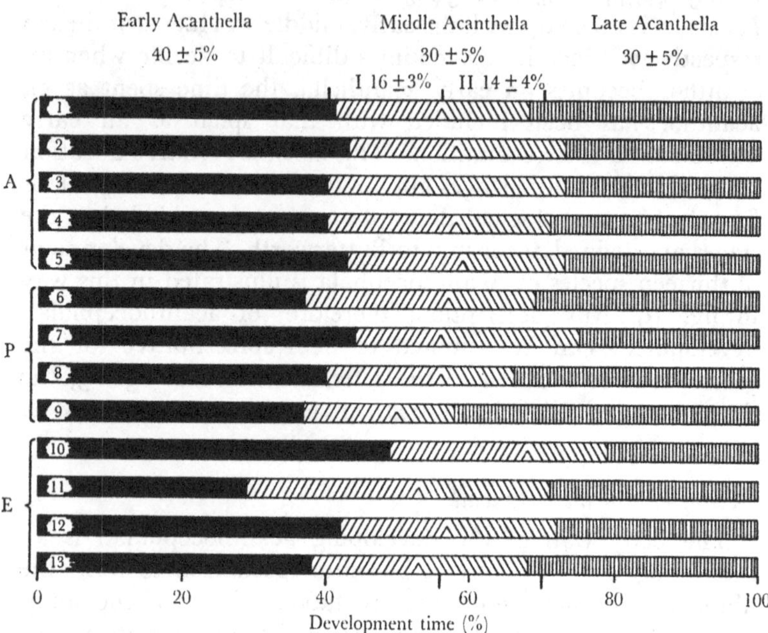

Fig. 19. Chart showing the nomenclature and percentage development time of the different stages of thirteen species of Acanthocephala. 1, *Macracanthorhynchus hirudinaceus* (Kates, 1943); 2, *Macracanthorhynchus ingens* (Moore, 1946*b*); 3, *Mediorhynchus grandis* (Moore, 1962); 4, *Moniliformis clarki* (Crook and Grundmann, 1964); 5, *Moniliformis dubius* (Moore, 1946*a*; King and Robinson, 1967); 6, *Echinorhynchus truttae* (Awachie, 1966); 7, *Leptorhynchoides thecatus* (DeGiusti, 1949*a*); 8, *Polymorphus minutus* (Butterworth, 1969*a*); 9, *Prosthorhynchus formosus* (Schmidt and Olsen, 1964); 10, *Neoechinorhynchus rutili* (Merritt and Pratt, 1964); 11, *Neoechinorhynchus emydis* (Hopp, 1954); 12, *Octospinifer macilentis* (Harms, 1965); 13, *Paulisentis fractus* (Cable & Dill, 1967). A, Archiacanthocephala; P, Palaeacanthocephala; E, Eoacanthocephala.

be the radial layer of the adult body wall (fig. 3*b*, p. 5). The first fibres of the felt layer appear in the region of the cortex where the organelles had congregated and electron dense material appears between the surface pores. In the late acanthella, the cuticle becomes visible and the felt layer increases in thickness in the part of the body wall which will contain the rest of the parasite after the cystacanth has formed. Butterworth found that the distance between the pores does not increase during development. Since the parasite's surface increases continuously, the pores must increase in number during development. The method of pore formation is unknown. This description of the development of the body wall of *P. minutus* can probably be applied to all acanthocephalans because their pattern of development is so consistent.

ASPECTS OF GROWTH AND METABOLISM DURING DEVELOPMENT

Most descriptions of the development of acanthocephalans include records of the external dimensions of the parasite and it is clear that the middle acanthella stage is usually the period of rapid elongation. *Polymorphus minutus* grows in length at that stage from about 0·05 mm to 2·0 mm; this is a 40-fold increase (Butterworth, 1969*b*). Measurements of length do not necessarily provide an accurate assessment of growth because of the invagination of the proboscis and other tissues. Butterworth therefore measured changes in dry weight. A graph constructed from Butterworth's data by plotting the dry weights of specimens of *P. minutus*, developing in *G. pulex*, against the age of the parasites is given in fig. 20. The curve, which is typical of that of any growing organism, shows that the most rapid phase of tissue synthesis, as opposed to increase in length, occurs during the late acanthella stage (fig. 19). Acanthellae increase in dry weight from 0·002 mg initially to 0·1 mg by the cystacanth stage (fig. 23).

Physiological investigations on organisms as small as developing acanthocephalans are difficult unless large numbers of parasites are available at the same stage of development; *Moniliformis dubius* and *M. hirudinaceus* are two convenient laboratory species for study, although little is known about their physiology during development. Some investigations have been undertaken on the metabolism of developing *P. minutus* by Butterworth (1968*b*, 1969*b*), who maintained this parasite *in vitro* in a saline adjusted to comply with some of the parasite's environmental conditions (table 12). It had already been found that acanthellae of *P. minutus*, which had been kept for a few hours in a similar salt solution, continued their development when returned by surgical procedures to an intermediate host's haemocoele (Crompton, 1967). Thus, Butterworth's results were obtained from parasites which were healthy throughout the experimental period.

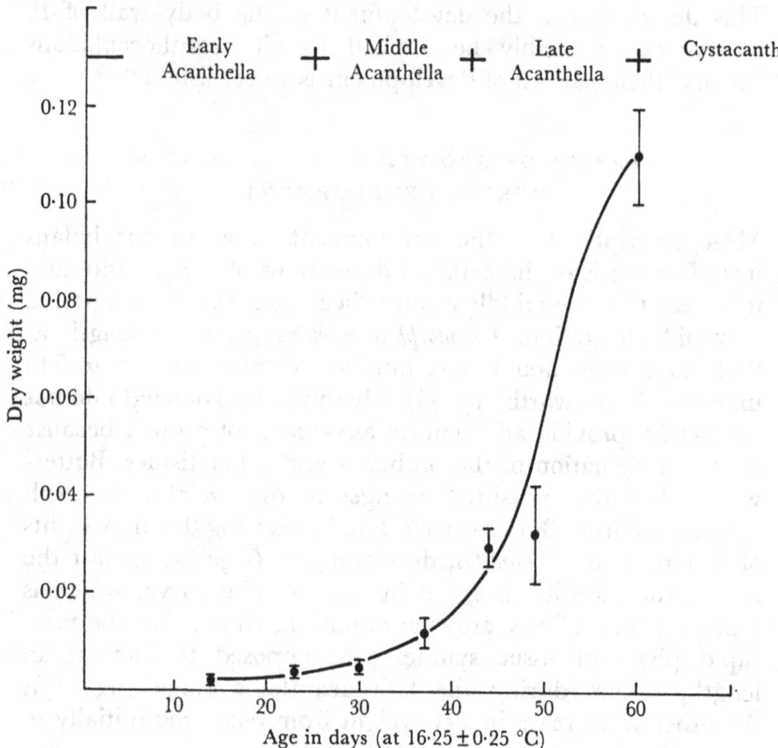

Fig. 20. The growth (dry weight) of *Polymorphus minutus* in its intermediate host *Gammarus pulex*. The amphipods were maintained at 16·25 ± 0·25 °C and the data were obtained from Butterworth (1969*b*) and P. E. Butterworth (personal communication).

From studies with radioactive glucose, it has been found that, for *P. minutus* kept at 13 °C, early acanthellae use 177 μg glucose/h per milligram of dry weight, middle acanthellae use 35 μg, late acanthellae use 16 μg and cystacanths use 2·10 μg, carbon dioxide being the only excretory product detected (Butterworth, 1969*b*). Since the cystacanth has an oxygen utilization rate of 0·102 ± 0·01 μl/h per milligram of dry weight and an R.Q. of 1·156, the developing parasite appears to metabolize carbohydrate for energy by processes involving oxygen. This tentative conclusion is supported by the fact that as little as 1·5% of the dry weight of cystacanths is glycogen.

The glycogen in developing acanthocephalans is not easily detected by histochemical methods until the parasites have become well-established, early acanthellae. For example, glycogen is not observed in *M. hirudinaceus* until about 25% of the developmental period has passed, when it is found to be associated with developing muscles and the body wall (Miller, 1943); similar results have been obtained by Crompton (1964*b*) for developing *P. minutus*. The presence of relatively little glycogen in early acanthellae may indicate that there is only enough glucose in the host's haemolymph to satisfy their high glucose requirement. Alternatively, the muscles and cortical tissues of the acanthellae may be insufficiently differentiated to permit glycogen storage.

Even less is known about lipid and protein metabolism in developing acanthocephalans although 28% of the dry weight of cystacanths of *P. minutus* is known to be lipid (Butterworth, 1968*b*). Over 95% of this lipid fraction is neutral lipid, most of which appears to be deposited in the radial layer of the body wall. In *P. minutus*, this lipid is coloured orange because it is associated with the carotenoid pigment esterified astaxanthin (Barrett and Butterworth, 1968). The significance of this pigment is not yet understood. The haemolymph of *G. pulex* contains five carotenoids irrespective of infections with *P. minutus*, but esterified astaxanthin is not a major pigment. *Polymorphus minutus*, therefore, may be orange because it selectively absorbs esterified astaxanthin from the host or because it converts any absorbed carotenoid to this pigment. Even so, no obvious function for the selective absorption of a carotenoid pigment can be proposed unless it forms part of a carotenoid–protein complex. In that case, the pigment could indicate the absorption of protein for amino acids from an environment where these compounds exist in low concentrations (table 12).

CHAPTER 8

Host–Parasite Reactions During the Development of Acanthocephala

TRADITIONAL definitions of parasitism usually assert that, during the course of an infection, an equilibrium develops between the resistance of the host and the infectiousness of the parasite. In this context, infectiousness is the capacity of a parasite to live in its host. The results from physiological studies of parasites, however, have led to interpretations of host–parasite relationships in terms of host biochemistry and parasite nutrition without recourse to antibodies, phagocytes and other agents. Thus, parasitism is considered to be the equilibrium in the host between substances which inhibit and those which promote growth of the parasite (Lewis, 1953). In this chapter, some aspects of the association between acanthocephalans and their intermediate hosts are discussed to show that both types of equilibrium occur and are integrated.

Most concepts of parasitism, however, are still in agreement that the association is of benefit to the parasite only and is always detrimental and often potentially lethal to the host. This generalization is also illustrated in this chapter by discussion of the effects of acanthocephalans on their intermediate hosts.

Within the haemocoele, insect haemocytes react by encapsulation against a variety of foreign bodies, but not against the insect's own healthy tissues, healthy intraspecific transplants and healthy natural parasites (Salt, 1961, 1963). The literature reveals that this generalization applies to most arthropods including amphipod crustaceans (fig. 21). Haemocytes, therefore, are an interesting factor in the environment of a developing acanthocephalan because they have the potential to resist or even destroy the parasite. They certainly react against and neutralize unhealthy acanthocephalans (Crompton, 1967). Some acanthocephalans can be seen to be enveloped in a thin transparent membrane during their association with the haemocoele. In fact, this membrane is a capsule produced by the activity of the haemocytes. A clue to the origin of these capsules may be obtained from the fact that acanthors of *Leptorhynchoides thecatus* do not move directly from the intestinal lumen to the haemocoele of the amphipod, *Hyalella azteca*, but remain and grow on the outer surface of the intestine (DeGiusti, 1949*a*). In this position the parasites, which are separated from the haemocytes by the thin serosa of connective tissue surrounding the intestine, appear to elicit a mild haemocytic reaction. Haemocytes are now known to be involved in the production of connective tissue in arthropods (Wigglesworth, 1956), and the haemocytic reaction observed by DeGiusti may have been concerned with restoration of the serosa, which had been stretched by the growing parasite, rather than with the parasite itself. On the basis of this supposition, evidence has been obtained to show that the prominent capsule surrounding *Polymorphus minutus*, during its development in the haemocoele of *Gammarus pulex*, originates from the wound-healing activity of the haemocytes involved in restoring the stretched serosa to its normal thickness (Crompton, 1964*a*). Eventually, the parasite becomes free from the outer surface of the intestine, but is enveloped in connective tissue which is extended by the haemocytes as the parasite's growth continues.

Evidence gleaned from the literature indicates that all the other acanthocephalans that have capsules could have acquired them in the manner described for *P. minutus* (Crompton, 1964*a*). Recently, aspects of the development of *Macracanthorhynchus ingens* (Bowen, 1967), *Prosthorhynchus formosus* (Schmidt and Olsen, 1964), *Moniliformis clarki* (Crook and Grundmann, 1964) and *Echinorhynchus truttae* (Awachie, 1966) have been described and their capsules can also be considered to have originated like that of *P. minutus*. Additional evidence for haemocytic participation in capsule formation has been provided by use of the electron microscope on *M. dubius* (Mercer and Nicholas, 1967) and *P. minutus* (Butterworth, 1969*a*). Their micrographs show that the capsules are composed of compact haemocytic tissue, but are different from those formed by haemocytes of the lepidopteran *Ephestia kuehniella* (Grimstone, Rotheram and Salt, 1967).

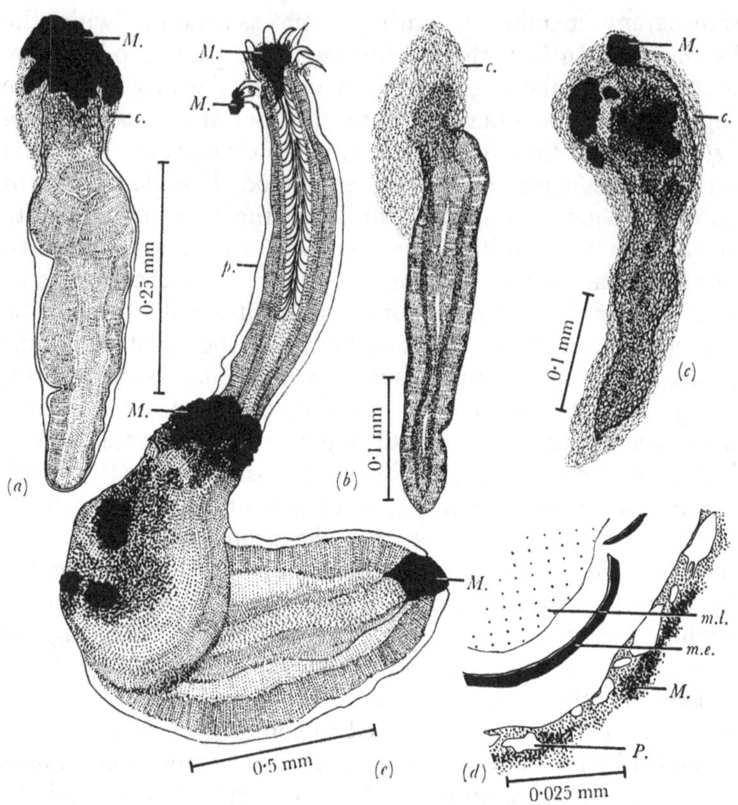

Fig. 21. The haemocytic reaction of *Gammarus*. (*a*) Acanthella of *Polymorphus minutus* in which the developing proboscis had stimulated the haemocytic reaction of *G. pulex*; (*b*) tip of hepatic caecum of *G. pulex* after 24 h in the haemocoele of another *G. pulex*. Note the cut end only has been encapsulated. (*c*) Tip of a hepatic caecum of *G. pulex* after 24 h in the haemocoele of *G. duebeni*. The whole organ has been encapsulated and melanin has been deposited at the cut end. (*d*) Diagrammatic representation of transverse section through an egg of *Aëdes aegypti* after 24 h in the haemocoele of *G. pulex*, showing the deposition of mucopolysaccharide and melanin; (*e*) capsule-free acanthella of *P. minutus* from *G. pulex* after 9 days in the haemocoele of another *G. pulex*. Note the pattern of melanin deposition. *c.*, Capsule; *m.e.*, mosquito egg; *m.l.*, mosquito larva; *p*, proboscis; *M.*, melanin; *P.*, mucopolysaccharide. (After Crompton, 1967, figs. 9, 11, 12, 17 and 18).

THE SIGNIFICANCE OF THE CAPSULE DURING DEVELOPMENT

Complications arise when attempts are made to explain the function and significance of capsules because some acanthocephalans develop without them. Examples include *Paulisentis fractus* from the copepod, *Tropocyclops prasinus* (Cable and Dill, 1967), and *Neoechinorhynchus* spp. from ostracods (Ward, 1940*a*; Hopp, 1954).

In the case of *P. minutus*, however, two possible functions for the capsule have been explored (Crompton, 1967). First, that it protects the delicate acanthellae from mechanical damage and, secondly, that it prevents the parasite from being encapsulated by its host's haemocytes. The second function may be postulated because the fact that the capsule is composed of healthy haemocytic tissue means that the haemocytes will not react against it (Salt, 1960). The surface of the developing acanthocephalan within the capsule will be foreign to the host's haemocytes which could be expected to react against it by encapsulation. This proposition has been tested by implanting fifteen acanthellae of *P. minutus*, from which the capsules had been removed by dissection, in the haemocoeles of uninfected *G. pulex*. Technical problems arose during the surgery and although none of the capsule-free parasites developed, none was encapsulated until it had become moribund (fig. 21*e*). It could be argued that transplantation of such a large object into a small amphipod could have damaged the parasite's surface and committed most of the available haemocytes to healing the cuticular wound. Clearly, a better test of the capsule as an inhibitor of encapsulation would have been to obtain acanthors of *P. minutus* and inject them directly into the amphipod's haemocoele. Recently, this experiment has been performed with *M. dubius* which usually develops within a capsule. Acanthors of this parasite were placed in the haemocoele of *Periplaneta americana* by King and Robinson (1967). These authors were not concerned with haemocytes or capsules and simply report that cystacanths developed when the infection was established by artificially avoiding the cockroaches' intestine. Thus, possession of a capsule, formed during an acanthocephalan's association with its host's intestinal tissues, does not appear to be necessary for the protection of a healthy acanthocephalan from destructive encapsulation.

Although no function may be apparent for capsules, their presence or absence may be tentatively interpreted as indicating the evolutionary state of the host–parasite relationship. This suggestion may be illustrated by comparing the penetration of the host's intestine by acanthors of *L. thecatus* and *P. minutus* on the one hand, and by acanthors of *Pomphorhynchus bulbocolli* and *P. fractus* on the other. *Leptorhynchoides thecatus* and *P. minutus* represent acanthocephalans which are associated for several days with their hosts' intestine and, in the case of *P. minutus*, a capsule is known to be produced by the time the parasite reaches the haemocoele (p. 79). *Pomphorhynchus bulbocolli*, which develops in the same intermediate host as *L. thecatus*, penetrates the host's intestine within a few hours (Jensen, 1952) and there is, therefore, insufficient time for a capsule of the type surrounding *P. minutus* to be formed. Jensen notes, however, that haemocytes repair the serosa damaged by the penetrating acanthors. *Paulisentis fractus* also reaches its host's haemocoele a few hours after ingestion of the eggs (Cable and Dill, 1967), but it is not known whether the host's haemocytes respond to the penetrating acanthors or not. Thus, *L. thecatus*, *P. minutus* and any encapsulated acanthocephalans appear to experience more host resistance than *P. bulbocolli*, *P. fractus* and other unencapsulated parasites. Greater host resistance is more likely to be a feature of relatively recent host–parasite relationships than of long-established ones.

This attempt to define the significance of the capsule has revealed that the establishment of an acanthocephalan infection is often resisted within the intestinal tissues of the intermediate host. Further support for this conclusion is available from several sources. When the amphipod, *H. azteca*, is maintained at 13 °C, eggs of *L. thecatus* hatch in the intestinal lumen, but the acanthors die in the intestinal tissues (DeGiusti, 1949*a*). In this case, the haemocytes appear to be able to repair the stretched serosa more rapidly than the acanthors can grow. *Macracanthorhynchus hirudinaceus* is able to reach the haemocoele of its host, *Popillia japonica*, irrespective of the temperature, but many acanthors always perish in the intestinal wall (Miller, 1943). Acanthors of *Mediorhynchus grandis* die in the intestinal wall of the dock beetle, *Gastroidea cyanea*, although they successfully reach the haemocoele of the orthopteroid insect, *Arphia luteola*, maintained at the same temperature (Moore, 1962). *Echinorhynchus coregoni* does not develop in *H. azteca* because the acanthors die in the amphipod's intestinal wall (DeGiusti, 1949*b*).

The best documented case of the death of acanthocephalans in their intermediate hosts is that recorded for *P. minutus* by

Hynes and Nicholas (1958). These authors discovered three strains of *P. minutus*; one strain is adapted to development in *G. pulex*, one in *G. lacustris* and one in *G. duebeni*. Thus, when eggs of *P. minutus* of the '*pulex*' strain are fed to *G. duebeni*, infections are rarely established and *vice versa*. The parasites die in the intestinal tissues apparently without direct contact with haemocytes. The osmotic pressure of the haemolymph of *G. duebeni* is higher than that of *G. pulex*, but this factor has been eliminated from being the cause of death of *P. minutus* of the '*pulex*' strain in *G. duebeni* (Crompton, 1964*b*).

No explanation is available at present to account for the deaths of these acanthocephalans, but there seems to be no justification for invoking humoral reactions since there is no evidence that they act against metazoan parasites in arthropods (Briggs, 1964). One feature, however, seems to be common to all the examples discussed. The acanthors perish unless they grow more rapidly than the serosa, as it is reinforced by haemocytes. In the case of *L. thecatus*, lowering the temperature affects the parasite more than the host and it appears, therefore, in this and the other cases, that the balance for the parasite between growth promoting and growth inhibiting substances in intestinal tissues favours inhibition (Lewis, 1953). The haemocytic response to penetrating parasites, and the production of capsules round some of them, is a form of host resistance which is not covered directly by Lewis' hypothesis. The success of the haemocytic response, however, appears to depend on the conditions in the intestinal tissues tending to retard the growth of the parasite. Thus, the conventional form of host resistance involving defensive agents and the more recent concept of resistance involving host biochemistry and parasite nutrition may both be present and associated in the relationships between acanthocephalans and their intermediate hosts.

SOME EFFECTS OF DEVELOPING ACANTHOCEPHALANS ON THEIR INTERMEDIATE HOSTS

Parasitic castration

Certain developing acanthocephalans are known to cause castration of their intermediate hosts (Reinhard, 1956). For example, *P. minutus* sterilizes mature female *G. pulex* and prevents immature females from reaching maturity (Le Roux, 1931*a*).

These effects are characterized by the disappearance of the setae on the oostegites, the cessation of ovarian development and the termination of mating behaviour. The same signs and symptoms can be produced in female *G. pulex* which have been irradiated with radium (Le Roux, 1931*b*). It is now known that reproduction in amphipods is under hormonal control (Steele, 1967), so that the absorption of female sex hormones by *P. minutus* could explain the observed effects. The parasite appears to have no effect on the sexual development of male *G. pulex*. By sterilizing female *Gammarus*, *P. minutus* acts in nature as a population regulator. Isolated populations of *G. lacustris*, in lakes, may be seriously diminished or eliminated by *P. minutus* (Hynes, 1955) and even hosts which colonize streams, for example, *G. pulex*, can be greatly reduced in numbers (Crompton and Harrison, 1965).

Another, as yet unidentified, acanthocephalan has been reported to interfere with the sexual development of the isopod, *Asellus aquaticus* (Munro, 1953). Twenty-eight intersexes were found in a sample of eighty infected isopods; the intersexes appeared to be modified female isopods since they possessed oostegites with immature setae.

Not all developing acanthocephalans castrate their hosts and *L. thecatus* does not appear to have any effects on the reproduction of the amphipod, *H. azteca* (Spaeth, 1951). Irradiation of female *H. azteca* with radium, however, causes the same effects as those reported by Le Roux (1931*b*) for *G. pulex*.

Other effects

The effects of developing acanthocephalans on the nutritional state of their hosts can be illustrated by considering an individual cystacanth of *P. minutus* in the haemocoele of *G. pulex*. A host weighing about 50 mg has a haemolymph volume of 12 μl, containing about 2·5 μg glucose and 0·33 μl dissolved oxygen at any one time (table 12, p. 78). A cystacanth metabolizes about 0·21 μg glucose/h and 0·01 μl oxygen/h *in vitro* under conditions designed to represent those in the haemocoele of *G. pulex* (Butterworth, 1969*b*). Consequently, an infected amphipod will have to eat more food and ventilate more vigorously if it is to supply the parasite's needs and maintain the normal glucose and oxygen levels of its haemolymph.

Effects of this type may not be noticed when intermediate hosts are maintained in a laboratory under conditions which avoid natural environmental pressures. Thus, there are reports of as many as thirty-one individuals of *P. minutus* developing in *G. pulex* in the laboratory (Crompton, 1964*b*), and of over 300 *M. hirudinaceus* in *P. japonica* (Miller, 1943). Recently, however, evidence has been obtained to show that acanthocephalans do weaken their hosts in nature. Movements of *G. pulex* upstream and drifting downstream are features of strong mature amphipods, but individuals infected with *Echinorhynchus truttae* are not able to move upstream and their attempts to do so cause them to be swept out of the population by the water current (Lehmann, 1967). This finding can be explained by assuming that the metabolism of the parasite strains the host beyond the point where it can obtain enough food for itself, the parasite and movement upstream. Thus, hosts will be unable to move upstream, and will be more easily caught by trout which form the final host of *E. truttae*.

CHAPTER 9

The Cystacanth Stage and the Infection of the Final Host

A cystacanth is defined here as the stage which has completed its structural development in the intermediate host, but still depends on the metabolism of that host for survival. The term 'cystacanth' is synonymous with 'infective acanthella' or 'juvenile' in some of the literature.

The cystacanth forms an intriguing stage in the life cycle of an acanthocephalan, because it must be adapted to endure adverse conditions such as gastric hydrochloric acid and high proteolytic activity, and yet respond to the stimuli which initiate the establishment of the worm in its correct environment. Cystacanths, therefore, present the parasitologist with an array of biological problems and it will be apparent from the discussion in this chapter that much remains to be discovered about their physiology.

Many references have been made to transport hosts in which immature acanthocephalans have been found encapsulated in the abdominal tissues. The significance of these hosts and their effects on acanthocephalans are also discussed in this chapter.

MORPHOLOGY AND PHYSIOLOGY OF CYSTACANTHS

Morphology

The cystacanth stage is reached when the proboscis, and often other tissues, are withdrawn into a portion of the parasite's body wall adapted to protect the inverted tissues from injury. Some cystacanths, for example, those of *Polymorphus minutus*, exhibit more invagination than others, for example, *Octospinifer macilentis* (fig. 22). The degree of invagination can probably be correlated with the amount of mechanical pressure and grinding activity to which cystacanths are exposed during their passage along the final host's alimentary tract. *Octospinifer macilentis*, which develops in ostracods, is a parasite of *Catostomus commersoni*, a stomachless fish which sucks its food off the surface of mud. Sufficient protection for cystacanths of *O. macilentis* is probably provided by the valves of the ostracod. *Polymorphus minutus*, in contrast, must withstand not only the squeezing action of a duck's bill, but also pressures of about 178 mm Hg generated as the ventriculus contracts four times/min (Farner, 1960). The nature of the food also affects the grinding activity of a duck's ventriculus; the harder the food, the stronger the contractions. The available evidence suggests that, in nature, *P. minutus* infects young Mallard and ducklings more easily than older birds (Crompton, 1969). This tendency could result from the preference of young Mallard for animal food (Millais, 1901) or from a greater loss of cystacanths through increased grinding in the ventriculus of older Mallard, which prefer seeds and vegetation.

Cystacanths of *Moniliformis dubius* appear to be intermediate in their degree of invagination when compared with those of *P. minutus* and *O. macilentis* (fig. 22). *Moniliformis dubius* may escape mechanical injury without being greatly invaginated because its cystacanths are small compared with the intermediate host, and perhaps smaller than particles usually swallowed by a rat. The cystacanths may become enmeshed in bits of tough cockroach cuticle and protected thereby.

Hardness, which partially depends on the degree of invagination, helps to provide protection from mechanical damage. In the case of *P. minutus*, hardness is achieved by the available space within the cystacanth being occupied by invaginated organs and viscous, oily liquid (Crompton, 1964*b*) and by special adaptations of the uninvaginated body wall. The development of this feature of the body wall begins with the migration of cortical nuclei just before the proboscis is inverted. The fibres of the felt layer then increase in number and it reaches a thickness of 20 μm (Butterworth, 1969*a*); in adult worms, the thickness decreases to 15 μm (Crompton, 1963). In cystacanths, the fibres are arranged in a layer of circular fibres surrounded on each side by a layer of longitudinal fibres (fig. 23). This arrangement is also characteristic of recently established worms, but changes later to the adult condition.

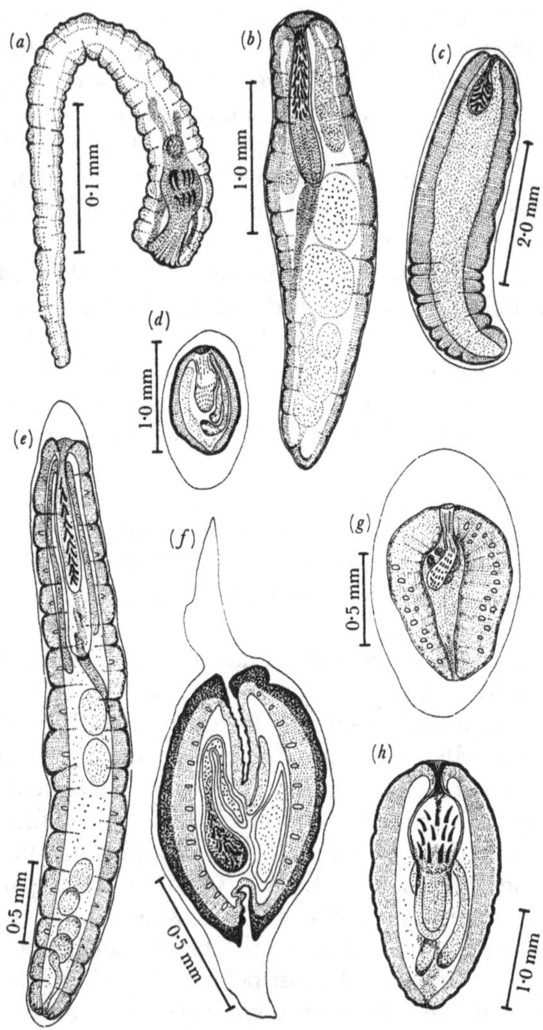

Fig. 22. Cystacanths of eight species of Acanthocephala. (*a*) *Octospinifer macilentis* (after Harms, 1965; fig. 15); (*b*) *Acanthocephalus ranae*; (*c*) *Macracanthorhynchus ingens* (after Moore, 1946*b*; fig. 9); (*d*) *Mediorhynchus grandis* (after Moore, 1962; fig. 11); (*e*) *Echinorhynchus truttae*; (*f*) *Polymorphus minutus*; (*g*) *Moniliformis dubius* (after Moore, 1946*a*; fig. 9); (*h*) *Macracanthorhynchus hirudinaceus* (after Kates, 1943, fig. 9).

Physiology

The cystacanth is a resting stage in the life cycle during the period between the end of development in the intermediate host and the parasite's arrival in the alimentary tract of the final or transport host. Little information is available about the retention of infectivity of cystacanths, but those of *P. minutus* have been reported to remain infective as long as their intermediate hosts live (Hynes and Nicholas, 1963). The normal life span of *Gammarus* is about 15 months (Hynes, 1955), but amphipods infected with *P. minutus* have been reported to live

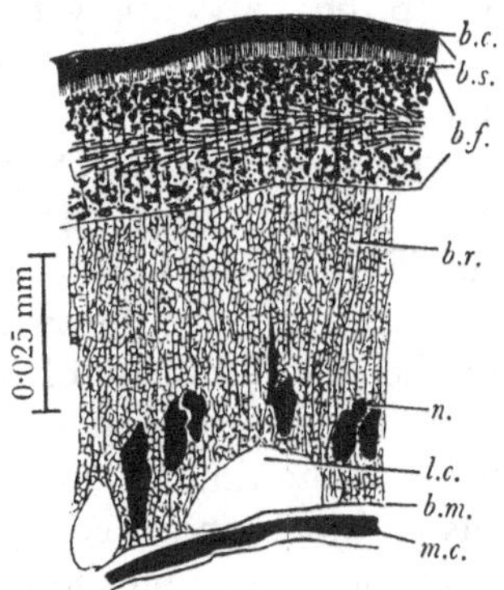

Fig. 23. Transverse section through the body wall of a cystacanth of *Polymorphus minutus*. *b.c.*, Cuticle; *b.f.*, felt layer; *b.m.*, basement membrane; *b.r.*, radial layer; *b.s.*, striped layer; *l.c.*, lacunar channel; *m.c.*, circular muscle; *n.*, nucleus.

up to 2 months longer than uninfected individuals from the same population (Hynes and Nicholas, 1963). *Gammarus* infected with *Echinorhynchus truttae*, in contrast, do not live as long as uninfected amphipods (Awachie, 1966).

The presence of much lipid in the tissues of cystacanths supports the conclusion that they are resting stages (Butterworth, 1969*b*). Over 90% of the lipid in cystacanths of *P. minutus* is cholesterol ester and it follows that the parasites must selectively absorb this substance from the host's haemolymph or synthesize it from other lipid products. There has been no demonstration yet that lipid in the cystacanth of *P. minutus* is used as an energy reserve during the resting period, but it is difficult to imagine its having any other function. Glycogen might also be expected to be an important metabolite in cystacanths, particularly since histochemical procedures have revealed prominent deposits in the radial layer of the body walls of *Macracanthorhynchus hirudinaceus* (Miller, 1943) and *P. minutus* (Crompton, 1964*b*). Cystacanths of *M. dubius*, however, have been found to contain relatively little glycogen (Graff and Kitzman, 1965) and only 1·5 % of the dry weight of cystacanths of *P. minutus* is glycogen (Butterworth, 1969*a*).

INFECTION OF THE FINAL HOST

Some of the factors stressed in connexion with the infection of intermediate hosts (p. 66), also apply to the infection of final hosts. The period of food retention by the stomach or analogous organ of the final host is a factor of major importance since it probably governs the length of time available for cystacanths to receive the activating stimuli without being irretrievably damaged by over-exposure to adverse conditions. Similarly, the rate of food passage through the alimentary tract also influences the establishment of an infection since the time taken for food to be propelled from the stomach to the worm's environment is the time available for an activated cystacanth to evert its proboscis and become attached.

Preliminary work has been done on the process of infection of ducks by *P. minutus* (Crompton, 1964*b*, 1969) and of rats by *M. dubius* (Graff and Kitzman, 1965). When ducks, whose intestines have reached the adult length of 150 cm, are starved for 16 h before being given cystacanths of *P. minutus* in *G. pulex*, worms can be found attached in their environment 50 to 100 min after being ingested. Such parasites have withstood a temperature change of at least 20 °C as well as the varying chemical and mechanical factors prevailing in 100 cm of the duck's alimentary tract. In a similar experiment involving 7-day-old ducklings, which have an intestinal length of 80 cm, parasites have been found attached in their environment after 35 to 70 min; these parasites have been propelled 50 cm. When ducks are allowed to feed for half an hour before eating the infected amphipods, no cystacanths have been observed beyond the ventriculus after 80 min. The results obtained from infections of rats by *M. dubius* have shown that more than half the number of cystacanths eaten by a rat are established as worms in their environment between 45 and 90 min later. (Graff and Kitzman 1965).

The process of infection in poikilothermic hosts is likely to take longer than it does in homeotherms. Thus, *O. macilentis* does not become attached in its environment until 4 h after being eaten with its ostracod host by *Catostomus commersoni* (Harms, 1965). Cystacanths of *E. truttae* remain in the trout's stomach for 20 h before everting their proboscides in the pyloric region of the intestine (Awachie, 1966).

The activation of cystacanths

Many parasitologists have attributed to host bile salts the function of activating normal parasites during the infection process. For example, bile salts have been found to initiate the evagination of cestode scolices (Rothman, 1959; Smyth, 1962). It is not surprising, therefore, that bile salts have been assumed to be involved in the activation of acanthocephalan cystacanths. The implication, already made in this chapter, that cystacanths are activated while in the stomach or ventriculus of their hosts still caters for the participation of bile salts since bile has been reported to be regurgitated from the duodenum to the anterior part of the alimentary tract (Groebbels, 1930). The possible correlation between bile and the activation of cystacanths should be interpreted with caution because cystacanths of *P. minutus* with everted proboscides have been found in the haemocoele of *Gammarus* (Hynes and Nicholas, 1958; Crompton, 1967).

A detailed study of the effects of bile salts on cystacanths of *M. dubius in vitro* has been made by Graff and Kitzman (1965). When cystacanths are incubated at 37 °C in saline containing bile salts, no eversion of the proboscis, which is the obvious sign of activation, occurs unless the concentration of the bile salt is greater than 0·005 %. Sodium taurocholate, at a concentration of 0·05 % and buffered to pH 8·5, activates 80 % of a group of cystacanths within 12 min. This bile salt is always more effective *in vitro* than sodium cholate and sodium glycocholate. Incubation of cystacanths in pepsin and trypsin has a questionable effect on activation, but the presence of CO_2 and reduction in oxygen tension enhances it. Conditions similar to those employed by Graff and Kitzman *in vitro* exist in the rat's alimentary tract and it is probable, therefore, that bile salts are involved in activation *in vivo*. Support for this conclusion is provided by the finding that, when the junctions of the bile ducts and intestines of rats are transferred by surgical procedures to the apices of the rats' caeca, no worms are recovered after each rat has eaten twenty cystacanths. Worms do become established, however, in rats which have experienced a sham operation. Once *M. dubius* is established in a rat, interference with the junction of the bile duct does not appear to affect the growth of the worm.

Thus, the evidence indicates that bile salts are involved in the activation of cystacanths *in vitro* and, since bile is involved in the establishment of infections in the host, these salts probably have a similar function *in vivo*. Bile salts are surface active agents which may alter the permeability of certain membranes within cystacanths and so stimulate the eversion of the proboscides. The fact that cystacanths are activated by conditions in the alimentary tracts of transport hosts indicates that bile salts may not have a part in determining host specificity for acanthocephalans as has been suggested for certain cestodes (Smyth, 1962).

CYSTACANTHS AND TRANSPORT HOSTS

Transport hosts, which are usually poikilothermic vertebrates, are those in which immature acanthocephalans are found encapsulated in the mesenteries, peritoneal tissues or liver (fig. 24). These hosts are of special interest because their intestinal tracts must provide adequate activation stimuli for cystacanths, but unsuitable conditions for growth and reproduction of adult worms. The presence of the acanthocephalans in the abdominal tissues suggests that the parasites react against the unfavourable intestinal environment by migrating away from it.

The infection of transport hosts

The process of infection must be similar to that operating when final hosts ingest cystacanths contained in their intermediate hosts, but the conditions which cause the activated parasite to leave the intestine are unknown. The assumption that adverse environmental factors are responsible for the invasion of the transport host's abdomen cannot be applied in all cases. For example, cystacanths of *Leptorhynchoides thecatus*, capable of establishing normal populations of the parasite in the final host's intestine, develop after 32 days in the intermediate host. If 30-day-old parasites are swallowed by the final host, the cystacanths are activated normally, but then bore through the intestinal wall and become encapsulated in the mesenteries (DeGiusti, 1939). More recently, specimens of *P. paradoxus* have been found living in the intestine and

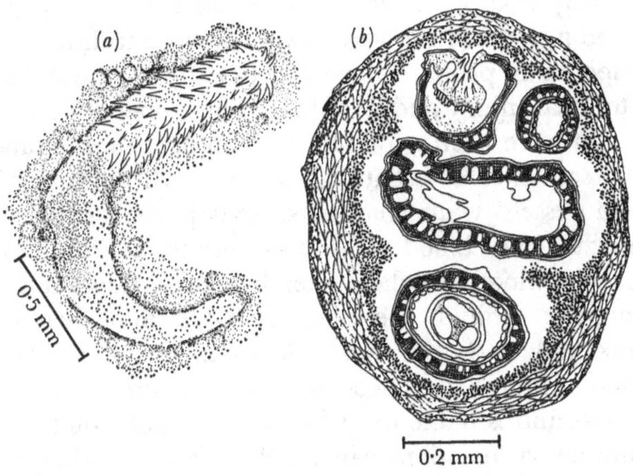

Fig. 24. The appearance of acanthocephalans in fish transport hosts. (*a*) *Leptorhynchoides thecatus* in the peritoneal tissues of yellow perch (after Van Cleave, 1920*a*; fig. 3); (*b*) transverse section through *Pallisentis basiri* in the liver of *Trichogaster chuna*. Note the encapsulating liver cells (after Hasan and Qasim, 1960, fig. 1).

encapsulated in the mesenteries of beavers, *Castor canadensis* (Connell and Corner, 1957), and specimens of *Eocollis arcanus* have been recovered from the intestines and livers of *Lepomis macrochirus* (Meade and Harvey, 1968). Final hosts are likely to feed on intermediate hosts containing acanthocephalans at different stages of development, and invasion of the abdomen by incompletely developed parasites may be a common feature of acanthocephalan biology. Such an event, however, may be terminal for the parasites unless they develop further to a state leading to normal infections if cannibalism is prevalent amongst final hosts; weasels, *Mustela itatsi itatsi*, have been reported to acquire *Gordiorhynchus itatsinis* in this manner (Kanda, 1958).

The environment of acanthocephalans in transport hosts

The capsule surrounding *Neoechinorhynchus cylindratus* in the liver of fish, *Lepomis* spp., is composed to two layers (Bogitsh, 1961). The inner layer appears to contain host connective tissue secreted by the outer layer of cells which are not recognizable as typical liver cells. These layers are clearly visible in the capsules produced in the liver of *Trichogaster chuna* against *Pallisentis basiri* (fig. 24*b*).

The encapsulation of acanthocephalans in transport hosts follows the pattern of the vertebrate response to inflammation. This response is primarily the reaction of an animal's blood vessels to local injury and irritation. Thus, once liver tissues or mesenteries have been penetrated by an acanthocephalan, the following sequence of events takes place. The permeability of the blood vessels' walls changes so that plasma exudes round the parasite and blood cells accumulate at the points of leakage. The blood circulation tends to slow down and leucocytes migrate through the capillary walls and surround the acanthocephalan. Later, host macrophages and other cells follow and reinforce the leucocytes. The parasite is now enclosed in host cells and is liable to suffer the effects of wound healing.

Inflammatory reactions can be classified according to their duration (Anderson and McCutcheon, 1966); Acanthocephala in transport hosts are associated with the chronic type. Once the parasites are established in the abdominal tissues, mild irritation will maintain the inflammatory response and tend to delay the neutralizing effect of wound healing. The fact that connective tissue is produced around *N. cylindratus* in its transport host (Bogitsh, 1961) signifies that wound healing, involving fibroblasts, eventually begins. Vertebrate fibroblasts synthesize collagen for connective tissue provided that the oxygen tension is above 10 mm Hg (Hunt, Zederfeldt and Dunphy, 1968). Thus acanthocephalans in transport hosts will be bathed in a fluid similar in chemical composition to blood plasma except that the oxygen content will have been reduced by the fibroblasts and other encapsulating cells. Furthermore, the impaired circulatory system in the region of inflammation will cause a rise in the tension of carbon dioxide. Consequently, in transport hosts, acanthocephalans inhabit an environment similar in some respects to the final host's small intestine;

survival is possible because the parasites are adapted to obtain energy independently of oxygen.

Various observations indicate that acanthocephalans survive for a limited time in transport hosts. Gradually, host resistance, in the form of wound healing, results in their death (Meyer, 1933). Degenerate specimens of *Pomphorhynchus bulbocolli* have been seen encapsulated in the mesenteries of fish (Ward, 1940*b*), and calcified *P. paradoxus* have been found in the mesenteries of beavers (Connell and Corner, 1957). It has been claimed that acanthocephalans of the genus *Centrorhynchus*, actually develop in the mesenteries of snakes (Gupta, 1950), but this report can be interpreted in terms of parasites found at differing degrees of degeneration in a transport host.

The significance of transport hosts in acanthocephalan life cycles

Baer (1961) stated that transport hosts are generally indispensable for the completion of the life cycles with which they are associated. Experimental support for this generalization is limited and most of the acanthocephalans cited in table 1 (p. 8) require intermediate and final hosts only.

In certain cases, however, transport hosts must be necessary elements in the life cycle. Some direct evidence for this conclusion is provided by considering *N. cylindratus* and circumstantial evidence may be obtained from discussion of *Corynosoma semerme* and *P. laevis*. The final host of *N. cylindratus* is the predacious bass, *Huro salmoides*, which, in nature, does not eat the ostracod intermediate host. The bluegill, *Lepomis pallidus*, does eat infected ostracods in the laboratory with the result that encapsulated parasites are found in its abdominal tissues (Ward, 1940*a*). When such bluegills are eaten by bass, mature *N. cylindratus* can be recovered from the intestine. Thus, *L. pallidus* appears to be an essential transport host for *N. cylindratus*.

Corynosoma semerme lives as an adult in the intestine of seals and, since seals feed on fish and not the amphipod intermediate host, *Pontoporeia affinis*, fish in the seals' diet must be transport hosts. Similarly, several of the predatory fish, which are final hosts of *P. laevis*, must become infected by eating the transport host, *Gasterosteus aculeatus*, rather than the intermediate host, *G. pulex*. The possibility also exists that seals, and the final hosts of *P. laevis*, become infected through the presence of undigested, infected amphipods in the stomachs of transport hosts rather than the presence of encapsulated parasites in their abdominal tissues.

In spite of the fact that relatively few transport hosts appear to be essential, it is obvious that many of them will assist the dispersal of the parasites, probably enable the parasites to survive conditions which would kill the intermediate hosts and possibly promote the evolution of new host–parasite relationships. For example, *Prosthorhynchus formosus* is normally a parasite of the isopod, *Armadillidium vulgare,* and passerine birds, but recently it has been found encapsulated in the mesenteries of the short-tailed shrew, *Blarina brevicauda* (Nickol and Oetinger, 1968). Shrews form the prey of owls and hawks and may eventually be the agent through which these birds become normal hosts of *P. formosus.*

Conclusion

It is evident from the preceding chapters, in which more than forty problems have been exposed, that much research remains to be done if the relationships between acanthocephalans and their hosts are to be understood. This research may be considered in three basic categories, which are:

(1) reinvestigations of problems already studied;

(2) investigations stimulated by earlier work;

(3) investigations into unknown aspects of acanthocephalan biology.

Research of the first category is necessary because acanthocephalans have been studied *in vitro* under conditions they never experience in their environments. Thus, the host–parasite relationship has been disturbed by experimental intervention and many of the results probably have little or no genuine significance. For example, although *Polymorphus minutus* almost certainly acquires glucose from its environment by the same mechanism as that operating *in vitro* (Crompton and Lockwood, 1968), no reliance should be placed on the rates of glucose uptake measured *in vitro* since the evidence, discussed in chapter 4, shows that glucose in the worm's environment should not be considered in isolation from oxygen, carbon dioxide and amino acids. Clearly, physiological problems of this type cannot be studied in the living, unrestricted host; instead, cultivation and growth of acanthocephalans *in vitro*, under conditions like those in the environment, must be achieved.

In some cases, the cultivation of parasites *in vitro* will be essential if research of the second category is to be attempted. For example, expansion of the studies on lipid uptake by *Acanthocephalus ranae* (Hammond, 1968*b*) can now be undertaken, either with this or other species of Acanthocephala, but the results will be of questionable value unless successful cultivation *in vitro* has been established. In other cases, investigations to elucidate problems arising from earlier work can be started without recourse to cultivation *in vitro*, although this facility will eventually be required. In the meantime, the effect of host sex hormones and host diet on the growth of acanthocephalans and the production of eggs, and intra- and interspecific competition in the environment should be investigated.

Research of the third category is necessary to provide information about life cycles, attachment positions, copulation, locomotion, osmoregulation, excretion, neurosecretion, diurnal rhythms, immunological reactions and numerous aspects of development in intermediate hosts. Several of these subjects will also be better understood if acanthocephalans are successfully cultivated *in vitro*.

The appeal of these brief concluding remarks is that major efforts should be made to cultivate acanthocephalans *in vitro*. In the case of parasites, this is definitely an ecological subject since cultivation *in vitro* is more likely to be achieved if more information is obtained about the parasites' environments.

References

ABBOTT, R. L. (1926). Contributions to the physiology of digestion in the Australian roach, *Periplaneta australasiae* Fab. *J. exp. Zool.* **44**, 219–53.

AL-HUSSAINI, A. H. (1949*a*). On the functional morphology of the alimentary tract of some fish in relation to differences in their feeding habits: anatomy and histology. *Q. Jl microsc. Sci.* **90**, 109–39.

AL-HUSSAINI, A. H. (1949*b*). On the functional morphology of the alimentary tract of some fish in relation to differences in their feeding habits: cytology and physiology. *Q. Jl microsc. Sci.* **90**, 323–54.

ALLEE, W. C., EMMERSON, A. E., PARK, O., PARK, T. and SCHMIDT, K. P. (1949). *Principles of Animal Ecology.* Philadelphia and London: W. B. Saunders Co.

ALLEN, E. A. (1936). *Tyzzeria perniciosa* gen. et sp. nov., a coccidium from the small intestine of the Pekin duck, *Anas domesticus* L. *Arch. Protistenk.* **87**, 262–66.

ALVAREZ, W. C. (1939). *An Introduction to Gastro-enterology.* New York and London: Paul B. Hoeber Inc.

ANDERSON, W. A. D. and MCCUTCHEON, M. (1966). Inflammation. In *Pathology*, vol. I, ed. W. A. D. Anderson. St. Louis: The C. V. Mosby Company.

ARME, C. and READ, C. P. (1969). Fluxes of amino acids between the rat and a cestode symbiote. *Comp. Biochem. Physiol.* **29**, 1135–47.

AWACHIE, J. B. E. (1966). The development and life history of *Echinorhynchus truttae* Schrank 1788, (Acanthocephala). *J. Helminth.* **40**, 11–32.

AWACHIE, J. B. E. (1967). Experimental studies on some host–parasite relationships of Acanthocephala. Co-invasion of *Gammarus pulex* L. by *Echinorhynchus truttae* Schrank, 1788 and *Polymorphus minutus* (Goeze, 1782). *Acta parasit. pol.* **15**, 69–74.

BAER, J. C. (1961). Embranchement des Acanthocéphales. In *Traité de Zoologie*, vol. IV, édité P-P. Grassé. Paris: Masson et Cie.

BALL, G. (1928). An acanthocephalid, *Corynosoma strumosum* (Rudolphi), in the California harbor seal. *Anat. Rec.* **41**, 82.

BARKER, J. and MACALISTER, A. (1865). On *Echinorhynchus porrigens. Proc. nat. Hist. Soc. Dublin* **4**, 293–94.

BARRETT, J. and BUTTERWORTH, P. E. (1968). The carotenoids of *Polymorphus minutus* (Acanthocephala) and its intermediate host, *Gammarus pulex. Comp. Biochem. Physiol.* **27**, 575–81.

BARRINGTON, E. J. W. (1942). Gastric digestion in the lower vertebrates. *Biol. Rev.* **17**, 1–27.

BARRINGTON, E. J. W. (1957). The alimentary canal and digestion. In *The Physiology of Fishes*, vol. I, ed. M. E. Brown. New York and London: Academic Press.

BAYLIS, H. A. (1931). *Gammarus pulex* as an intermediate host for trout parasites. *A. Mag. nat. Hist.* 10th series **7**, 431–5.

BAYLIS, H. A. (1944). Three new Acanthocephala from marine fishes of Australasia. *A. Mag. nat. Hist.* 11th series **11**, 462–72.

BEAMES, C. G. and FISHER, F. M. (1964). A study on the neutral lipids and phospholipids of the Acanthocephala *Macracanthorhynchus hirudinaecus* and *Moniliformis dubius*. *Comp. Biochem. Physiol.* **13**, 401–12.

BECKER, E. R. (1959). Protozoa. In *Diseases of Poultry*, 4th edition, ed. H. E. Biester and L. H. Schwarte. Ames, Iowa: Iowa State University Press.

BENTON, A. H. (1954). Notes on *Moniliformis clarki* (Ward) in Eastern New York (Moniliformidae: Acanthocephala). *J. Parasit.* **40**, 102–3.

BEZUBIK, B. (1957). Studies on *Polymorphus minutus* (Goeze, 1782)–syn. *Polymorphus magnus* Skrjabin, 1913. *Acta parasit. pol.* **5**, 1–8.

BHALERAO, G. D. (1931). On a new species of Acanthocephala from *Ophicephalus striatus*. *A. Mag. nat. Hist.* 10th series, **7**, 569–73.

BOGITSH, B. (1961). Histological and histochemical observations on the nature of the cyst of *Neoechinorhynchus cylindratus* in *Lepomis* sp. *Proc. helminth. Soc. Wash.* **28**, 75–81.

BOWEN, R. C. (1967). Defense reactions of certain spirobolid millipedes to larval *Macracanthorhynchus ingens*. *J. Parasit.* **53**, 1092–5.

BRAND, T. von (1939*a*). Chemical and morphological observations upon the composition of *Macracanthorhynchus hirudinaceus*. (Acanthocephala). *J. Parasit.* **25**, 329–42.

BRAND, T. von (1939*b*). The glycogen distribution in the body of Acanthocephala. *J. Parasit.* **25**, suppl. 22.

BRAND, T. von (1940). Further observations on the composition of Acanthocephala. *J. Parasit.* **26**, 301–7.

BRAND, T. von (1966). *Biochemistry of Parasites*. New York and London: Academic Press.

BRIGGS, J. D. (1964). Immunological responses. In *The Physiology of Insecta*, vol. III, ed. M. Rockstein. New York and London: Academic Press.

BRYANT, C. and NICHOLAS, W. L. (1965). Intermediary metabolism in *Moniliformis dubius* (Acanthocephala). *Comp. Biochem. Physiol.* **15**, 103–12.

BRYANT, C. and NICHOLAS, W. L. (1966). Studies on the oxidative metabolism of *Moniliformis dubius* (Acanthocephala). *Comp. Biochem. Physiol.* **17**, 825–40.

BUCK, J. B. (1953). Physical properties and chemical composition of insect blood. In *Insect Physiology*, ed. K. D. Roeder. London: Chapman and Hall Ltd.

BULLOCK, W. L. (1949*a*). Histochemical studies on the Acanthocephala. I. The distribution of lipase and phosphatase. *J. Morph.* **84**, 185–200.

BULLOCK, W. L. (1949*b*). Histochemical studies on the Acanthocephala. II. The distribution of glycogen and fatty substances. *J. Morph.* **84**, 201–26.

BULLOCK, W. L. (1958). Histochemical studies on the Acanthocephala. III. Comparative histochemistry of alkaline glycerophosphatase. *Expl Parasit.* **7**, 51–68.

BULLOCK, W. L. (1960). Histochemical studies on the Acanthocephala. IV. Acid phosphatase distribution in some Neoechinorhynchidae. *Sobretiro del libro Homenaje al Doctor Eduardo Caballero y Caballero*. Mexico. 423–8.

Bullock, W. L. (1962). A new species of *Acanthocephalus* from New England fishes, with observations on variability. *J. Parasit.* **48**, 442–51.

Bullock, W. L. (1963). Intestinal histology of some salmonid fishes with particular reference to the histopathology of acanthocephalan infections. *J. Morph.* **112**, 23–44.

Bullock, W. L. (1967). The intestinal histology of the mosquito fish, *Gambusia affinis* (Baird and Girard). *Acta zool. Stockh.* **48**, 1–17.

Burlingame, P. L. and Chandler, A. C. (1941). Host-parasite relations of *Moniliformis dubius* (Acanthocephala) in albino rats, and the environmental nature of resistance to single and superimposed infections with this parasite. *Am. J. Hyg.* **33**, 1–21.

Butterworth, P. E. (1968*a*). An estimation of the haemolymph volume in *Gammarus pulex. Comp. Biochem. Physiol.* **26**, 1123–5.

Butterworth, P. E. (1968*b*). Aspects of the nutrition of *Polymorphus minutus* during its development. *Parasitology* **58**, 3P.

Butterworth, P. E. (1969*a*). The development of the body wall of *Polymorphus minutus* (Acanthocephala) in the intermediate host, *Gammarus pulex. Parasitology* **59**, 373–88.

Butterworth, P. E. (1969*b*). *Studies on the Physiology and Development of* Polymorphus minutus. Ph.D. Dissertation, University of Cambridge.

Byrd, E. E. and Denton, F. J. (1949). The helminth parasites of birds. II. A new species of Acanthocephala from North American birds. *J. Parasit.* **35**, 391–410.

Cable, R. M. and Dill, W. T. (1967). The morphology and life history of *Paulisentis fractus* Van Cleave and Bangham, 1949 (Acanthocephala: Neoechinorhynchidae). *J. Parasit.* **53**, 810–17.

Chaicharn, A. and Bullock, W. L. (1967). The histopathology of acanthocephalan infections in suckers with observations on the intestinal histology of two species of catostomid fishes. *Acta zool. Stockh.* **48**, 19–42.

Chance, B., Schoener, B. and Schindler, F. (1964). The Intracellular Oxidation–Reduction State. In *Oxygen in the Animal Organism.* I.U.B Symposium, 31, ed. F. Dickens and E. Neil. Oxford. London. Edinburgh. New York. Paris. Frankfurt: Pergamon Press.

Chino, H. and Gilbert, L. I. (1965). Lipid release and transport in insects. *Biochim. Biophys. Acta.* **98**, 94–110.

Chubb, J. C. (1964). Occurrence of *Echinorhynchus clavula* (Dujardin, 1845) nec Hamann, 1892 (Acanthocephala) in the fish of Llyn Tegid (Bala Lake), Merionethshire. *J. Parasit.* **50**, 52–9.

Chubb, J. C. (1965). Mass occurrence of *Pomphorhynchus laevis* (Müller, 1776), Monticelli 1905 (Acanthocephala) in the chub *Squalius cephalus* (L) from the river Avon, Hampshire. *Parasitology* **55**, 5P.

Connell, R. and Corner, A. H. (1957). *Polymorphus paradoxus* sp. nov. (Acanthocephala) parasitizing beavers and muskrats in Alberta, Canada. *Can. J. Zool.* **35**, 525–33.

Cornwell, P. B. (1968). *The Cockroach,* vol. I. London: Hutchinson.

Crompton, D. W. T. (1963). Morphological and histochemical observations on *Polymorphus minutus* (Goeze, 1782), with special reference to the body wall. *Parasitology* **53**, 663–85.

CROMPTON, D. W. T. (1964*a*). The envelope surrounding *Polymorphus minutus* (Goeze, 1782) (Acanthocephala) during its development in the intermediate host, *Gammarus pulex*. *Parasitology* **54**, 721–35.

CROMPTON, D. W. T. (1964*b*). *Studies on Acanthocephala, with special reference to* Polymorphus minutus. Ph.D. Dissertation, University of Cambridge.

CROMPTON, D. W. T. (1965). A histochemical study of the distribution of glycogen and oxidoreductase activity in *Polymorphus minutus* (Goeze, 1782) (Acanthocephala). *Parasitology* **55**, 503–14.

CROMPTON, D. W. T. (1966). Measurements of glucose and amino acid concentrations, temperature and pH in the habitat of *Polymorphus minutus* (Acanthocephala) in the intestine of domestic ducks. *J. exp. Biol.* **45**, 279–84.

CROMPTON, D. W. T. (1967). Studies on the haemocytic reaction of *Gammarus* spp., and its relationship to *Polymorphus minutus* (Acanthocephala). *Parasitology* **57**, 389–401.

CROMPTON, D. W. T. (1969). On the environment of *Polymorphus minutus* (Acanthocephala) in ducks. *Parasitology* **58**, 19–28.

CROMPTON, D. W. T. and EDMONDS, S. J. (1969). Measurements of the osmotic pressure in the habitat of *Polymorphus minutus* (Acanthocephala) in the intestine of domestic ducks. *J. exp. Biol.* **50**, 69–77.

CROMPTON, D. W. T. and HARRISON, J. G. (1965). Observations on *Polymorphus minutus* (Goeze, 1782) (Acanthocephala) from a wildfowl reserve in Kent. *Parasitology* **55**, 345–55.

CROMPTON, D. W. T. and LEE, D. L. (1965). The fine structure of the body wall of *Polymorphus minutus* (Goeze, 1782) (Acanthocephala). *Parasitology* **55**, 357–64.

CROMPTON, D. W. T. and LOCKWOOD, A. P. M. (1968). Studies on the absorption and metabolism of D-(u-^{14}C) glucose by *Polymorphus minutus* (Acanthocephala) *in vitro*. *J. exp. Biol.* **48**, 411–25.

CROMPTON, D. W. T. and NESHEIM, M. C. (1969*a*). Amino acid patterns in the small intestine. *Proc. Nutr. Soc.* **28**, 20A.

CROMPTON, D. W. T. and NESHEIM, M. C. (1969*b*). Amino acid patterns during digestion in the small intestine of ducks. *J. Nutr.* **99**, 43–50.

CROMPTON, D. W. T. and NESHEIM, M. C. (1970). Lipid, bile acid, water and dry matter content of the intestinal tract of domestic ducks with reference to the habitat of *Polymorphus minutus* (Acanthocephala). *J. exp. Biol.* **52**, 437–55.

CROMPTON, D. W. T., SHRIMPTON, D. H. and SILVER, I. A. (1965). Measurements of the oxygen tension in the lumen of the small intestine of the domestic duck. *J. exp. Biol.* **43**, 473–8.

CROMPTON, D. W. T. and WARD, P. F. V. (1967). Lactic and succinic acids as excretory products of *Polymorphus minutus* (Acanthocephala) *in vitro*. *J. exp. Biol.* **46**, 423–30.

CROMPTON, D. W. T. and WHITFIELD, P. J. (1968*a*). A hypothesis to account for the anterior migrations of *Hymenolepis diminuta* (Cestoda) and *Moniliformis dubius* (Acanthocephala) in the intestine of rats. *Parasitology* **58**, 227–9.

CROMPTON, D. W. T. and WHITFIELD, P. J. (1968*b*). The course of infection and egg production of *Polymorphus minutus* (Acanthocephala) in domestic ducks. *Parasitology* **58**, 231–46.

CROOK, J. R. and GRUNDMANN, A. W. (1964). The life history and larval development of *Moniliformis clarki* (Ward, 1917). *J. Parasit.* **50**, 689–93.

CROSS, S. X. (1934). A probable case of non-specific immunity between two parasites of ciscoes of the trout lake region of Northern Wisconsin. *J. Parasit.* **20**, 244–5.

DAWES, B. (1929). The histology of the alimentary tract of the plaice (*Pleuronectes platessa*). *Q. Jl microsc. Sci.* **73**, 243–74.

DAWES, B. (1930). The absorption of fats and lipoids in the plaice (*P. platessa* L.). *J. mar. biol. Ass. U.K.* **17**, 75–102.

DAY, M. F. (1952). Wound healing in the gut of the cockroach *Periplaneta*. *Aust. J. scient. Res.* B **5**, 282–9.

DAY, M. F. and POWNING, R. F. (1949). A study of the processes of digestion in certain insects. *Aust. J. scient. Res.* B **2**, 175–215.

DAY, M. F. and WATERHOUSE, D. F. (1953). Structure of the alimentary system. In *Insect Physiology*, ed. K. D. Roeder. London: Chapman and Hall Ltd.

DECHTIAR, A. O. (1968). *Neoechinorhynchus carpiodi* n. sp. (Acanthocephala: Neoechinorhynchidae) from quillback of Lake Erie. *Can. J. Zool.* **46**, 201–4.

DEGIUSTI, D. L. (1939). Further studies on the life cycle of *Leptorhynchoides thecatus*. *J. Parasit.* **25**, suppl., 22.

DEGIUSTI, D. L. (1949*a*). The life cycle of *Leptorhynchoides thecatus* (Linton), an acanthocephalan of fish. *J. Parasit.* **35**, 437–60.

DEGIUSTI, D. L. (1949*b*). Partial development of *Echinorhynchus coregoni* in *Hyalella azteca* and the cellular reaction of the amphipod to the parasite. *J. Parasit.* **35**, suppl., 31.

DEGIUSTI, D. L., BEIGELMAN, E. and DELIDOW, S. (1962). A comparison of the pH values of the anatomical areas of the digestive tract of the amphipods *Hyalella azteca* (Saussure) and *Gammarus limnaeus* (Smith). *Trans. Am. microsc. Soc.* **81**, 262–4.

DENNY, M. (1968). The life cycle and ecology of *Polymorphus marilis* Van Cleave, 1939. *Parasitology* **58**, 23 P.

DIXON, M. and WEBB, E. C. (1964). *Enzymes*. 2nd edition. London: Longmans, Green and Co. Ltd.

DOGIEL, V. A. (1961). Ecology of the parasites of freshwater fishes. In *Parasitology of Fishes*, ed. V. A. Dogiel, G. K. Petrushevski and Y. I. Polyanski. Translated by Z. Kabata. Edinburgh and London: Oliver and Boyd.

DUNAGAN, T. T. (1962). Studies on *in vitro* survival of Acanthocephala. *Proc. helminth. Soc. Wash.* **29**, 131–5.

DUNAGAN, T. T. (1964). Studies on the carbohydrate metabolism of *Neoechinorhynchus* spp. (Acanthocephala). *Proc. helminth. Soc. Wash.* **31**, 166–72.

DUNAGAN, T. T. and SCHEIFINGER, C. C. (1966*a*). Studies on the TCA cycle of *Macracanthorhynchus hirudinaceus* (Acanthocephala). *Comp. Biochem. Physiol.* **18**, 663–7.

DUNAGAN, T. T. and SCHEIFINGER, C. C. (1966*b*). Studies on glycolytic enzymes from *Macracanthorhynchus hirudinaceus* (Acanthocephala). *J. Parasit.* **52**, 730–4.

Edmonds, S. J. (1965). Some experiments on the nutrition of *Moniliformis dubius* Meyer (Acanthocephala). *Parasitology* **55**, 337–44.

Edmonds, S. J. (1966). Hatching of the eggs of *Moniliformis dubius*. *Expl Parasit.* **19**, 216–26.

Edmonds, S. J. and Dixon, B. R. (1966). Uptake of small particles by *Moniliformis dubius* (Acanthocephala). *Nature, Lond.* **209**, 99.

Ekbaum, E. (1938). Notes on the occurrence of Acanthocephala in Pacific fishes. I. *Echinorhynchus gadi* (Zoega) Müller in salmon and *E. lageniformis* sp. nov. and *Corynosoma strumosum* (Rudolphi) in two species of flounder. *Parasitology* **30**, 267–73.

Fairbairn, D. (1958). Trehalose and glucose in helminths and other invertebrates *Can. J. Zool.* **36**, 787–96.

Fairbairn, D., Wertheim, G., Harper, R. P. and Schiller, E. L. (1961). Biochemistry of normal and irradiated strains of *Hymenolepis diminuta*. *Expl Parasit.* **11**, 248–63.

Farner, D. S. (1960). Digestion and the digestive system. In *Biology and Comparative Physiology of Birds*, vol. I, ed. A. J. Marshall. New York and London: Academic Press.

Fenn, W. O. (1964). Introduction. In *Oxygen in the Animal Organism*. I.U.B. Symposium, 31, ed. F. Dickens and E. Neil. Oxford. London. Edinburgh. New York. Paris. Frankfurt: Pergamon Press.

Fisher, F. M. (1964). Synthesis of trehalose in Acanthocephala. *J. Parasit.* **50**, 803–4.

Florkin, M. (1960). Blood chemistry. In *The Physiology of Crustacea*, vol. I, ed. T. H. Waterman. New York and London: Academic Press.

Florkin, M. and Jeuniaux, C. (1964). Hemolymph: Composition. In *The Physiology of Insects*, vol. III, ed. M. Rockstein. New York and London: Academic Press.

Follansbee, R. (1945). The osmotic activity of gastrointestinal fluids after water ingestion in the rat. *Am. J. Physiol.* **144**, 355–62.

Forster, G. R. (1953). Peritrophic membranes in the Caridea (Crustacea: Decapods). *J. mar. biol. Ass. U.K.* **32**, 315–18.

Garden, E. A., Rayski, C. A., and Thom, V. M. (1964). A parasitic disease in Eider Ducks. *Bird Study* **11**, 280–7.

Gauld, D. T. (1957). A peritrophic membrane in calanoid copepods. *Nature, Lond.* **179**, 325–6.

Gilmour, D. (1965). *The Metabolism of Insects*. Edinburgh and London: Oliver and Boyd.

Ginetsinskaya, T. A. (1961). The life cycles of fish helminths and the biology of their larval stages. In *Parasitology of Fishes*, ed. V. A. Dogiel, G. K. Petrushevski and Y. I. Polyanski. Translated by Z. Kabata. Edinburgh and London: Oliver and Boyd.

Glasgow, R. D. and Deporte, P. (1939). Recovery from excreta of the pigeon of viable eggs of the giant thorny-headed worm of swine. *J. econ. Ent.* **32**, 882.

Glocklin, V. C. and Fairbairn, D. (1952). The metabolism of *Heterakis gallinae*. I. Aerobic and anaerobic respiration: carbohydrate sparing action of carbon dioxide. *J. cell. comp. Physiol.* **39**, 341–56.

Golvan, Y. (1962). Acanthocéphales parasites des mammifères. *Helminth. Abstr.* **33**, 210.

GOODRICH, H. P. (1928). Reactions of *Gammarus* to injury and disease, with notes on some microsporidial and fungoid diseases. *Q. Jl microsc. Sci.* **72**, 325–53.

GORDON, H. A. and BRUCHNER-KARDOSS, E. (1961). Effect of normal microbial flora on intestinal surface area. *Am. J. Physiol.* **201**, 175–8.

GRAFF, D. J. (1964). Metabolism of ^{14}C-glucose by *Moniliformis dubius* (Acanthocephala). *J. Parasit.* **50**, 230–4.

GRAFF, D. J. (1965). The utilization of $C^{14}O_2$ in the production of acid metabolites by *Moniliformis dubius* (Acanthocephala). *J. Parasit.* **51**, 72–5.

GRAFF, D. J. and ALLEN, K. (1963). Glycogen content in *Moniliformis dubius* (Acanthocephala). *J. Parasit.* **49**, 204–8.

GRAFF, D. J. and KITZMAN, W. B. (1965). Factors influencing the activation of acanthocephalan cystacanths. *J. Parasit.* **51**, 424–9.

GREEF, R. (1864). Untersuchungen über den Bau und die Naturgeschichte von *Echinorhynchus miliarius* Zenker (*E. polymorphus*). *Arch. Naturgesch.* **30**, 98–140.

GRIMSTONE, A. V., ROTHERAM, S. and SALT, G. (1967). An electron-microscope study of capsule formation by insect blood cells. *J. Cell Sci.* **2**, 281–92.

GROEBBELS, F. (1930). Die Verdauung der Hausgans, untersucht mit der Methode der Dauerkanüle. *Pflügers Arch. ges. Physiol.* **224**, 687–701.

GUPTA, P. V. (1950). On some stages in the development of the acanthocephalan genus *Centrorhynchus*. *Indian J. Helminth.* **2**, 41–8.

HAMMOND, R. A. (1966*a*). Changes of internal hydrostatic pressure and body shape in *Acanthocephalus ranae*. *J. exp. Biol.* **45**, 197–202.

HAMMOND, R. A. (1966*b*). The proboscis mechanism of *Acanthocephalus ranae*. *J. exp. Biol.* **45**, 203–13.

HAMMOND, R. A. (1967*a*). The fine structure of the trunk and praesoma wall of *Acanthocephalus ranae* (Schrank, 1788), Lühe, 1911. *Parasitology*, **57**, 475–86.

HAMMOND, R. A. (1967*b*). The mode of attachment within the host of *Acanthocephalus ranae* (Schrank, 1788), Lühe, 1911. *J. Helminth.* **41**, 321–8.

HAMMOND, R. A. (1968*a*). Observations on the body surface of some acanthocephalans. *Nature, Lond.* **218**, 872–3.

HAMMOND, R. A. (1968*b*). Some observations on the role of the body wall of *Acanthocephalus ranae* in lipid uptake. *J. exp. Biol.* **48**, 217–25.

HARMS, C. E. (1965). The life cycle and larval development of *Octospinifer macilentis* (Acanthocephala: Neoechinorhynchidae). *J. Parasit.* **51**, 286–93.

HASAN, R. and QASIM, S. Z. (1960). The occurrence of *Pallisentis basiri* Farooqi (Acanthocephala) in the liver of *Trichogaster chuna* (Ham.) *Z. Parasitkde* **20**, 152–6.

HASELWOOD, G. A. D. (1964). The biological significance of chemical differences in bile salts. *Biol. Rev.* **39**, 537–74.

HAWKER, L. E., LINTON, A. H., FOLKES, B. F. and CARLILE, M. J. (1960). *An Introduction to the Biology of Micro-organisms*. London: Edward Arnold Ltd.

HIBBARD, K. M. and CABLE, R. M. (1968). The uptake and metabolism of tritiated glucose, tyrosine and thymidine by adult *Paulisentis fractus* Van Cleave and Bangham, 1949 (Acanthocephala: Neoechinorhynchidae). *J. Parasit.* **54**, 517–23.

HOLMES, J. C. (1961). Effects of concurrent infections on *Hymenolepis diminuta* (Cestoda) and *Moniliformis dubius* (Acanthocephala). I. General effects and comparison with crowding. *J. Parasit.* **47**, 209–16.

HOLMES, J. C. (1962). Effects of concurrent infections on *Hymenolepis diminuta* (Cestoda) and *Moniliformis dubius* (Acanthocephala). III. Effects in hamsters. *J. Parasit.* **48**, 97–100.

HOPP, W. B. (1954). Studies on the morphology and life-cycle of *Neoechinorhynchus emydis* (Leidy), an acanthocephalan parasite of the map turtle, *Graptemys geographica* (La Sueur). *J. Parasit.* **40**, 284–99.

HUNT, T. K., ZEDERFELDT, B. and DUNPHY, J. E. (1968). Role of oxygen tension in healing. *Q. Jl surgical Sci.* **4**, 29–85.

HUZINGA, H. W. and HALEY, A. J. (1962). The occurrence of the acanthocephalan parasite, *Telosentis tenuicornis*, in the spot, *Leiostomus xanthurus*, in Chesapeake Bay. *Chesapeake Sci.* **3**, 35–42.

HYMAN, L. H. (1951). *The Invertebrates*, vol. III. New York, Toronto and London: McGraw-Hill Book Company, Inc.

HYNES, H. B. N. (1955). The reproductive cycle of some British freshwater *Gammaridae. J. Anim. Ecol.* **24**, 352–87.

HYNES, H. B. N. and NICHOLAS, W. L. (1957). The development of *Polymorphus minutus* (Goeze, 1782) (Acanthocephala) in the intermediate host. *Ann. trop. Med. Parasit.* **51**, 380–91.

HYNES, H. B. N. and NICHOLAS, W. L. (1958). The resistance of *Gammarus* spp. to infection by *Polymorphus minutus* (Goeze, 1782) (Acanthocephala). *Ann. trop. Med. Parasit.* **52**, 376–83.

HYNES, H. B. N. and NICHOLAS, W. L. (1963). The importance of the acanthocephalan *Polymorphus minutus* as a parasite of domestic ducks in the United Kingdom. *J. Helminth.* **37**, 185–98.

INOUE, I. (1967). *Eimeria saitamae* n. sp.: a new cause of coccidiosis in domestic ducks (*Anas platyrhyncha* var. domestica). *Jap. J. vet. Sci.* **29**, 209–15.

JENSEN, T. (1952). *The life cycle of the fish acanthocephalan,* Pomphorhynchus bulbocolli *(Linkins) Van Cleave 1919, with some observations on larval development* in vitro. Ph.D. Dissertation, University of Minnesota.

JOHNSTON, T. H. and DELAND, E. W. (1929). Australian Acanthocephala, No. 2. *Trans. R. Soc. S. Aust.* **53**, 155–66.

KANDA, Y. (1958). Studies on *Mustela itatsi itatsi* as natural intermediate and final hosts of *Gordiorhynchus itatsinis* (Acanthocephala) in Japan. *Jap. J. Parasit.* **7**, 589–98.

KATES, K. C. (1942). Viability of the eggs of the swine thorn-headed worm (*Macracanthorhynchus hirudinaceus*). *J. agric. Res.* **64**, 93–100.

KATES, K. C. (1943). Development of the swine thorn-headed worm, *Macracanthorhynchus hirudinaceus* in its intermediate host. *Am. J. vet Res.* **4**, 173–81.

KATES, K. C. (1944). Some observations on experimental infections of pigs with the thorn-headed worm, *Macracanthorhynchus hirudinaceus. Am. J. vet Res.* **5**, 166–72.

KILEJAN, A. (1963). The effect of carbon dioxide on glycogenesis in *Moniliformis dubius* (Acanthocephala). *J. Parasit.* **49**, 862–3.

KING, D. and ROBINSON, E. S. (1967). Aspects of the development of *Moniliformis dubius. J. Parasit.* **53**, 142–9.

KINNE, O. (1959). Ecological data on the amphipod, *Gammarus duebeni.* A monograph. *Veröff. Inst. Meeresforsch. Bremerh.* **6**, 177–202.

KOFOID, C. A., McNEIL, E. and CAILLEAU, R. (1932). Electrometric pH determinations of the walls and contents of the gastro-intestinal tracts of normal albino rats. *Univ. Calif. Publs Zool.* **36**, 347–55.

LAURIE, J. S. (1957). The *in vitro* fermentation of carbohydrates by two species of cestode and one species of Acanthocephala. *Expl Parasit.* **6**, 245–60.

LAURIE, J. S. (1959). Aerobic metabolism of *Moniliformis dubius* (Acanthocephala). *Expl Parasit.* **8**, 188–97.

LEEUWENHOEK, A. (1692). Letter to the Royal Society of London, written from Delft on 16th September 1692.

LEHMANN, U. (1967). Drift und Populationsdynamik von *Gammarus pulex fossarum* Koch. *Z. Morph. Ökol. Tiere* **60**, 227–74.

LEIBOWITZ, L. (1969). *Wenyonella philiplevinei*, n. sp., a coccidial organism of the white Pekin Duck. *Avian Dis.* **12**, 670–81.

LELOIR, L. F. and MUÑOZ, J. M. (1938). Ethyl alcohol metabolism in animal tissues. *Biochem. J.* **32**, 299–307.

LE ROUX, M. L. (1931*a*). Castration parasitaire et caractères sexuels secondaires chez les Gammariens. *C. r. hebd. Séanc. Acad. Sci., Paris* **192**, 889–91.

LE ROUX, M. L. (1931*b*). La castration expérimentale de femelles de Gammariens et sa répercussion sur l'évolution des oostegites. *C. r. hebd. Séanc. Acad. Sci., Paris* **193**, 885–7.

LEWIS, R. W. (1953). An outline of the balance hypothesis of parasitism. *Am. Nat.* **87**, 273–81.

LINNANE, A. W. (1965). Aspects of the biosynthesis of the mitochondria of *Saccharomyces cerevisiae.* In *Oxidases and Related Redox Systems*, 2, ed. T. E. King, H. S. Mason and M. Morrison. New York. London. Sydney: John Wiley and Sons Inc.

LOCKWOOD, A. P. M. (1961). The urine of *Gammarus duebeni* and *G. pulex. J. exp. Biol.* **38**, 647–58.

LOCKWOOD, A. P. M. (1968). *Aspects of the Physiology of Crustacea.* Edinburgh and London: Oliver and Boyd.

LYSTER, L. L. (1940). Parasites of some Canadian sea mammals. *Can. J. Res.* **18**, 395–409.

MANTER, H. W. (1928). Notes on the eggs and larvae of the thorny-headed worm of hogs. *Trans. Am. microsc. Soc.* **47**, 342–7.

MEADE, T. G. and HARVEY, J. S. (1968). New record of numbers and sites of infection in fishes by the acanthocephalan, *Eocollis arcanus* Van Cleave. *J. Parasit.* **54**, 371.

MERCER, E. H. and NICHOLAS, W. L. (1967). The ultrastructure of the capsule of the larval stages of *Moniliformis dubius* (Acanthocephala) in the cockroach, *Periplaneta americana. Parasitology* **57**, 169–74.

MERCER, E. H. and DAY, M. F. (1952). The fine structure of the peritrophic membrane of certain insects. *Biol. Bull. mar. biol. Lab., Woods Hole* **103**, 384–94.

Merritt, S. V. and Pratt, I. (1964). The life history of *Neoechinorhynchus rutili* and its development in the intermediate host (Acanthocephala: Neoechinorhynchidae). *J. Parasit.* **50**, 394–400.

Meyer, A. (1931). Urhautzelle, Hautbahn und plasmodiale Entwicklung der Larve von *Neoechinorhynchus rutili* (Acanthocephala). *Zool. Jb.* **53**, 103–26.

Meyer, A. (1933). *Acanthocephala. Bronn's Klassen und Ordnungen des Tierreichs*, 4, Leipzig. C. F. Winter'sche Verlagshandlung.

Millais, J. G. (1901). *The Natural History of the British Surface Feeding Ducks.* London, New York and Bombay: Longmans, Green and Co.

Miller, M. A. (1943). Studies on the developmental stages and glycogen metabolism of *Macracanthorhynchus hirudinaceus* in the Japanese beetle larva. *J. Morph.* **73**, 19–42.

Mistilis, S. P. and Birchall, A. (1969). Induction of alcohol dehydrogenase in the rat. *Nature, Lond.* **223**, 199–200.

Monné, L. (1959). On the external cuticles of various helminths and their role in the host–parasite relationship. *Ark. Zool.* **12**, 343–58.

Monné, L. (1964). Chemie und Bildung der Embryophoren von *Polymorphus botulus* Van Cleave (Acanthocephala). *Z. Parasitkde* **25**, 148–56.

Monné, L. and Hönig, G. (1954). On the embryonic envelopes of *Polymorphus botulus* and *P. minutus* (Acanthocephala). *Ark. Zool.* **7**, 257–60.

Moore, D. V. (1942). An improved technique for the study of the acanthor stage in certain acanthocephalan life histories. *J. Parasit.* **28**, 495.

Moore, D. V. (1946*a*). Studies on the life history and development of *Moniliformis dubius* Meyer, 1933. *J. Parasit.* **32**, 257–71.

Moore, D. V. (1946*b*). Studies on the life history and development of *Macracanthorhynchus ingens* Meyer, 1933, with a redescription of the adult worm. *J. Parasit.* **32**, 387–99.

Moore, D. V. (1962). Morphology, life history and development of the acanthocephalan *Mediorhynchus grandis* Van Cleave, 1916. *J. Parasit.* **48**, 76–86.

Mueller, J. F. (1929). Studies on the microscopical anatomy and physiology of *Ascaris lumbricoides* and *Ascaris megalocephala*. *Z. Zellforsch.* **8**, 362–403.

Munro, W. R. (1953). Intersexuality in *Asellus aquaticus* L. parasitized by a larval acanthocephalan. *Nature, Lond.* **172**, 313.

Nasset, E. S. (1968). Contribution of the digestive system to the amino acid pool. In *Protein Nutrition and Free Amino Acid Patterns*, ed. J. H. Leathem. New Brunswick, New Jersey: Rutgers University Press.

Nicholas, W. L. (1967). The biology of the Acanthocephala. In *Adv. Parasitol.* **5**, 205–46. New York and London: Academic Press.

Nicholas, W. L. and Grigg, H. (1965). The *in vitro* culture of *Moniliformis dubius* (Acanthocephala). *Expl Parasit.* **16**, 332–40.

Nicholas, W. L. and Hynes, H. B. N. (1958). Studies on *Polymorphus minutus* (Goeze, 1782) (Acanthocephala) as a parasite of the domestic duck. *Ann. trop. Med. Parasit.* **52**, 36–47.

Nicholas, W. L. and Hynes, H. B. N. (1963). The embryology of *Polymorphus minutus* (Acanthocephala). *Proc. zool. Soc. Lond.* **141**, 791–801.

Nicholas, W. L. and Mercer, E. H. (1965). The ultrastructure of the tegument of *Moniliformis dubius* (Acanthocephala). *Q. Jl microsc. Sci.* **106**, 137–46.

NICKOL, B. B. and OETINGER, D. F. (1968). *Prosthorhynchus formosus* from the short-tailed shrew (*Blarina brevicauda*) in New York State. *J. Parasit.* **54**, 456.

O'RIORDAN, A. M. (1968). *The absorption of electrolytes in the midgut of the cockroach.* Ph.D. Dissertation, University of Cambridge.

PARENTI, U., ANTONIOTTI, M. L. and BECCIO, C. (1965). Sex ratio and sex digamety in *Echinorhynchus truttae*. *Experientia* **21**, 657–8.

PARK, T. (1962). Beetles, competition and populations. *Science, N.Y.* **138**, 1369–75.

PETERS, W. (1969). Vergleichende Untersuchungen der Feinstruktur peritrophischer Membranen von Insekten. *Z. Morph. Tiere* **64**, 21–58.

PETROSCHENKO, V. I. (1956). *Acanthocephala of Domestic and Wild Animals.* I. (In Russian). Moskva: Akad. Nauk. SSSR.

PETROSCHENKO, V. I. (1958). *Acanthocephala of Domestic and Wild Animals.* II. (In Russian.) Moskva: Akad. Nauk. SSSR.

PFLUGFELDER, O. (1949). Histophysiologische Untersuchungen über die Fettresorption darmloser Parasiten: Die Funktion der Lemnisken der Acanthocephala. *Z. Parasitkde.* **14**, 274–80.

PODDER, T. N. (1937). A new species of Acanthocephala, *Neoechinorhynchus topseyi* n. sp., from a Calcutta fish, *Polynemus heptadactylus* (Cuv. & Val.) *Parasitology* **29**, 365–9.

PODDER, T. N. (1938). A new species of Acanthocephala, *Acanthosentis dattai* n. sp., from a freshwater fish of Bengal, *Barbus ticto* (Ham. & Buch). and *B. stigma* (Cuv. & Val.). *Parasitology*, **30**, 171–5.

PRAKASH, A. and ADAMS, J. R. (1960). A histopathological study of the intestinal lesions induced by *Echinorhynchus lageniformis* (Acanthocephala —Echinorhynchidae) in the starry founder. *Can. J. Zool.* **38**, 895–7.

PRATT, J. R. (1950). A qualitative analysis of the free amino acids in insect blood. *Ann. ent. Soc. Am.* **43**, 573–80.

RAUTHER, M. (1930). Acanthocephala = Kratzwürmer. In *Handbuch der Zoologie*, vol. II, 3–5, gegründet W. Kukenthal. Berlin und Leipzig: Walter de Gruyter and Co.

RAWSON, D. (1952). The occurrence of parasitic worms in British freshwater fishes. *A. Mag. nat. Hist.* 12th series **5**, 877–87.

READ, C. P. (1950). The vertebrate small intestine as an environment for parasitic helminths. *Rice Inst. Pamph.* **37**, no. 2, 1–94.

READ, C. P. and ROTHMAN, A. H. (1958). The carbohydrate requirement of *Moniliformis* (Acanthocephala). *Expl Parasit.* **7**, 191–7.

READ, C. P., ROTHMAN, A. H. and SIMMONS, J. E. (1963). Studies on membrane transport, with special reference to parasite–host integration. *Ann. N.Y. Acad. Sci.* **113**, 154–205.

REDI, F. (1684). *Osservazioni intorno agli animali viventi che si trovano negli animali viventi.* Firenze.

REINHARD, E. G. (1956). Parasitic castration of Crustacea. *Expl Parasit.* **5**, 79–107.

ROGERS, W. P. (1949). On the relative importance of aerobic metabolism in small nematode parasites of the alimentary tract. I. Oxygen tension in the normal environment of the parasites. *Aust. J. scient. Res.* B **2**, 166–74.

ROGERS, W. P. (1962). *The Nature of Parasitism.* New York and London: Academic Press.

ROBINSON, E. S. (1965). The chromosomes of *Moniliformis dubius* (Acanthocephala). *J. Parasit.* **51**, 430–2.

ROTHMAN, A. H. (1958). Role of bile salts in the biology of tapeworms. I. Effects on the metabolism of *Hymenolepis diminuta* and *Oochoristica symmetrica*. *Expl Parasit.* **7**, 328–37.

ROTHMAN, A. H. (1959). Studies on the excystment of tapeworms. *Expl Parasit.* **8**, 336–64.

ROTHMAN, A. H. (1967). Ultrastructural enzyme localization in the surface of *Moniliformis dubius* (Acanthocephala). *Expl Parasit.* **21**, 42–6.

ROTHMAN, A. H. and FISHER, F. M. (1964). Permeation of amino acids in *Moniliformis* and *Macracanthorhynchus* (Acanthocephala). *J. Parasit.* **50**, 410–14.

ROTHSCHILD, LORD (1961). *A Classification of Living Animals.* London: Longmans, Green and Co. Ltd.

SALT, G. (1960). Surface of a parasite and the haemocytic reaction of its host. *Nature, Lond.* **188**, 162–3.

SALT, G. (1961). The haemocytic reaction of insects to foreign bodies. In *The Cell and the Organism.* Cambridge University Press.

SALT, G. (1963). The defence reactions of insects to metazoan parasites. *Parasitology* **53**, 527–642.

SANDFORD, E. W. (1918). Experiments on the physiology of digestion in the blattidae. *J. exp. Zool.* **25**, 355–411.

SARS, G. (1867). *Histoire Naturelle des Crustacés D'eau Douce de Norvège. I. Les Malacostraces.* Christiana: Johnsen.

SCHAD, G. A. (1966). Immunity, competition and natural regulation of helminth populations. *Am. Nat.* **100**, 359–64.

SCHEIBEL, L. W. and SAZ, H. J. (1966). The pathway for anaerobic carbohydrate dissimilation in *Hymenolepis diminuta. Comp. Biochem. Physiol.* **18**, 151–62.

SCHMIDT, G. D. (1963). *Arhythmorhynchus capellae* sp. n. (Polymorphidae: Acanthocephala), a parasite of the common snipe *Capella gallinago delicata. J. Parasit.* **49**, 483–4.

SCHMIDT, G. D. (1965). *Corynosoma bipapillum* sp. n. from Bonaparte's gull *Larus philadelphia* in Alaska, with a note on *C. constrictum* Van Cleave, 1918. *J. Parasit.* **51**, 814–16.

SCHMIDT, G. D. and OLSEN, O. W. (1964). Life cycle and development of *Prosthorhynchus formosus* (Van Cleave, 1918) Travassos, 1926, an acanthocephalan parasite of birds. *J. Parasit.* **50**, 721–30.

SCHWARTZ, B. (1929). Important internal parasites of livestock. *Vet. Med.* **24**, 336–46.

SITA, E. (1949). The life cycle of *Moniliformis moniliformis* (Bremser, 1811), Acanthocephala. *Curr. Sci.* **18**, 216–18.

SMITH, M. H. (1969). The pigments of Nematoda and Acanthocephala. In *Chemical Zoology*, Vol. III, ed. M. Florkin and B. T. Scheer. New York and London: Academic Press.

SMYTH, J. D. (1962). Lysis of *Echinococcus granulosus* by surface-active agents in bile and the role of this phenomenon in determining host specificity in helminths. *Proc. Roy. Soc.* B **156**, 553–72.

SMYTH, J. D. and HASELWOOD, G. A. D. (1963). The biochemistry of bile as a factor in determining host specificity in intestinal parasites, with particular reference to *Echinococcus granulosus*. *Ann. N.Y. Acad. Sci.* **113**, 234–60.

SNIPES, B. T. and TAUBER, O. E. (1937). Time required for food passage through the alimentary tract of the cockroach, *Periplaneta americana* Linn. *Ann. ent. Soc. Am.* **30**, 277–84.

SPAETH, F. W. (1951). The influence of acanthocephalan parasites and radium emanations on the sexual characters of *Hyalella* (Crustacea: Amphipoda). *J. Morph.* **88**, 361–83.

SPENCER, R. P. (1960). *The Intestinal Tract*. Springfield, Illinois: C. C. Thomas.

SPINDLER, L. A. and KATES, K. C. (1940). Survival on soil of eggs of the swine thorn-headed worm *Macracanthorhynchus hirudinaceus*. *J. Parasit.* **26**, suppl. 19.

STEELE, V. J. (1967). Resting stage in the reproductive cycles of *Gammarus*. *Nature, Lond.* **214**, 1034.

STRANACK, F. R., WOODHOUSE, M. A. and GRIFFIN, R. L. (1966). Preliminary observations on the ultrastructure of the body wall of *Pomphorhynchus laevis* (Acanthocephala). *J. Helminth.* **40**, 395–402.

STUNKARD, H. W. (1965). New intermediate hosts in the life cycle of *Prosthenorchis elegans* (Diesing, 1851), an acanthocephalan parasite of primates. *J. Parasit.* **51**, 645–9.

STYCZYŃSKA, E. (1958). Some observations on the development and bionomics of larvae of *Filicollis anatis* Schrank. *Acta parasit. Pol.* **6**, 213–24.

SUTCLIFFE, D. W. (1963). The chemical composition of haemolymph in insects and some other arthropods, in relation to their phylogeny. *Comp. Biochem. Physiol.* **9**, 121–35.

TAUBER, O. E. (1937). The effect of ecdysis on the number of mitotically dividing cells in the hemolymph of the insect *Blatta orientalis*. *Ann. ent. Soc. Am.* **30**, 35–9.

TREHERNE, J. E. (1957). Glucose absorption in the cockroach. *J. exp. Biol.* **34**, 478–85.

THRELFALL, W. (1968). Helminth parasites of some birds in Newfoundland. *Can. J. Zool.* **46**, 909–13.

VAN CLEAVE, H. J. (1920*a*). Notes on the life cycle of two species of Acanthocephala from freshwater fishes. *J. Parasit.* **6**, 167–72.

VAN CLEAVE, H. J. (1920*b*). Acanthocephala parasitic in the dog. *J. Parasit.* **7**, 91–4.

VAN CLEAVE, H. J. (1923). *Telosentis*, a new genus of Acanthocephala from Southern Europe. *J. Parasit.* **9**, 174–5.

VAN CLEAVE, H. J. (1952). Some host–parasite relationships of the Acanthocephala, with special reference to the organs of attachment. *Expl Parasit.* **1**, 305–30.

VAN CLEAVE, H. J. and ROSS, E. L. (1944). Physiological responses of *Neoechinorhynchus emydis* (Acanthocephala) to various solutions. *J. Parasit.* **30**, 369–72.

VAN CLEAVE, H. J. and HADERLIE, E. C. (1950). A new species of the acanthocephalan genus *Octospinifer* from California. *J. Parasit.* **36**, 169–73.

VENARD, C. E. and WARFEL, J. H. (1953). Some effects of two species of

Acanthocephala on the alimentary canal of the large mouth bass. *J. Parasit.* **39**, 187–90.
Vonk, H. J. (1960). Digestion and metabolism. In *The Physiology of Crustacea*, vol. I, ed. T. H. Waterman. New York and London: Academic Press.
Ward, H. (1940*a*). Studies on the life-history of *Neoechinorhynchus cylindratus* (Van Cleave, 1913) (Acanthocephala). *Trans. Am. microsc. Soc.* **59**, 327–47.
Ward, H. B. (1940*b*). Notes on juvenile Acanthocephala. *J. Parasit.* **26**, 191–3.
Ward, P. F. V. and Crompton, D. W. T. (1969). The alcoholic fermentation of glucose by *Moniliformis dubius* (Acanthocephala) *in vitro*. *Proc. Roy. Soc.* B **172**, 65–88.
Waterhouse, D. F. and McKellar, J. W. (1961). The distribution of chitinase activity in the body of the American cockroach. *J. Insect Physiol.* **6**, 185–95.
Waterman, T. H. (1960). Comparative physiology. In *The Physiology of Crustacea*, vol. II, ed. T. H. Waterman. New York and London: Academic Press.
Weisel, G. F. (1962). Comparative study of the digestive tract of a sucker, *Catostomus catostomus*, and a predacious minnow, *Ptychocheilus oregonense*. *Am. Midl. Nat.* **68**, 334–46.
West, A. J. (1963). A preliminary investigation of the embryonic layers surrounding the acanthor of *Acanthocephalus jacksoni* Bullock, 1962 and *Echinorhynchus gadi* (Zoega) Müller, 1776. *J. Parasit.* **49**, suppl., 42–3.
West, A. J. (1964). The acanthor membranes of two species of Acanthocephala. *J. Parasit.* **50**, 731–4.
Wharton, D. R. A., Wharton, M. L. and Lola, J. E. (1965). Cellulase in the cockroach, with special reference to *Periplaneta americana* (L.). *J. Insect Physiol.* **11**, 947–59.
Wheeler, R. E. (1963). Studies on the total haemocyte count and haemolymph volume in *Periplaneta americana* (L.) with special reference to the last moulting cycle. *J. Insect Physiol.* **9**, 223–35.
Whitfield, P. J. (1968). A histological description of the uterine bell of *Polymorphus minutus* (Acanthocephala). *Parasitology* **58**, 671–82.
Whitfield, P. J. (1969). *Studies on the Reproduction of Acanthocephala*. Ph.D. Dissertation, University of Cambridge.
Wigglesworth, V. B. (1927*a*). Digestion in the cockroach. I. The hydrogen ion concentration in the alimentary canal. *Biochem. J.* **21**, 791–6.
Wigglesworth, V. B. (1927*b*). Digestion in the cockroach. II. The digestion of carbohydrates. *Biochem. J.* **21**, 797–811.
Wigglesworth, V. B. (1928). Digestion in the cockroach. III. The digestion of proteins and fats. *Biochem. J.* **22**, 150–61.
Wigglesworth, V. B. (1956). The haemocytes and connective tissue formation in an insect, *Rhodnius prolixus* (Hemiptera). *Q. Jl microsc. Sci.* **97**, 87–98.
Wigglesworth, V. B. (1959). Insect blood cells. *A. Rev. Ent.* **4**, 1–16.
Wilson, G. S. and Miles, A. A. (1964). *Topley and Wilson's Principles of Bacteriology and Immunity*. London: Edward Arnold Ltd.

WILSON, T. H. (1962). *Intestinal Absorption.* Philadelphia and London: W. B. Saunders Co.

WISEMAN, G. (1964). *Absorption from the Intestine.* New York and London: Academic Press.

WOLFFHÜGEL, K. (1908). Sobre *Echinorhynchus hirudinaceus* (Pall). *Revta Cent. Estud. Agron., B. Aires* 3–7.

WOLFFHÜGEL, K. (1924). Versuche mit dem Riesenkratzer (*Macracanthorhynchus hirudinaceus* (Pallas), Syn. *Echinorhynchus gigas* Goeze). *Z. InfektKrankh. parasit. Krankh. Hyg. Haustiere* **26**, 177–207.

WOLVEKAMP, H. P. and WATERMAN, T. H. (1960). Respiration. In *The Physiology of Crustacea*, vol. I, edited T. H. Waterman. New York and London: Academic Press.

WRIGHT, R. D. and LUMSDEN, R. D. (1968). Ultrastructural and histochemical properties of the acanthocephalan epicuticle. *J. Parasit.* **54**, 1111–23.

WYATT, G. R. (1961). The biochemistry of insect hemolymph. *A. Rev. Ent.* **6**, 75–102.

WYATT, G. R. (1967). The biochemistry of sugars and polysaccharides in insects. In *Adv. Insect Physiol.* **4**, 287–360. New York and London: Academic Press.

YAMAGUTI, S. (1935). Studies on the helminth fauna of Japan Part 8. Acanthocephala, I. *Jap. J. Zool.* **6**, 247–78.

YAMAGUTI, S. (1963). *Systema Helminthum.* V. *Acanthocephala.* New York and London: John Wiley and Sons.

YAMAGUTI, S. and MIYATA, I. (1942). Über die Entwicklungsgeschichte von *Moniliformis dubius* Meyer, 1933 (Acanthocephala) mit besonderer Berücksichtigung seiner Entwicklung im Zwischenwirt. 32 pages, published by authors from Parasitology Laboratory, University of Tokyo.

ZOPPI, G. and SHMERLING, D. H. (1969). Intestinal disaccharidase activities in some birds, reptiles and mammals. *Comp. Biochem. Physiol.* **29**, 289–94.

Index